Your guide to viewing
more than 150 sky wonders

OBSERVING WITH SMALL TELESCOPES

Kevin Ritschel

Kalmbach
Media

Kalmbach Media
21027 Crossroads Circle
Waukesha, Wisconsin 53186
www.MyScienceShop.com

Published in 2023
27 26 25 24 23   1 2 3 4 5

Manufactured in China

ISBN: 978-1-62700-924-9
EISBN: 978-1-62700-925-6

Cover Photo: Joel Short
Editor: Rich Talcott
Book Design: Lisa Schroeder
Sky Map Illustrations: Roen Kelly

Library of Congress Control Number: 2022945204

CONTENTS

INTRODUCTION

I wrote this book for many reasons. I wrote this book to help people see the cosmos with their own eyes, to be amazed at our universe, and to appreciate the beauty of the night sky.

I wrote this book to show that you don't need a huge and expensive telescope to see a lot of the wonderful objects you've undoubtedly seen in breathtaking images. Yes, a long-exposure image with a space telescope or huge observatory instrument can show you lots more detail, and in "living color," but I think you will appreciate those efforts even more if you have looked at the skies with your own eyes. You'll also find it doesn't cost a fortune to do this.

I wrote this book to give all the recipients of "Christmas telescopes" some direction on what to look for in the sky with them. If you pursue observing the sky with a telescope, I think you will also soon appreciate that a $79 Christmas special isn't going to allow you to find many objects in the sky because of a wobbly mount and inadequate finder scope. But I will give you tips on what to spend some money on to get rewarding vistas.

I wrote this book to dispel the notion that you need a lot of power or magnification to see anything in the night sky. You don't. In fact, too much power and not enough dark, non-light-polluted skies foil most attempts to see the cosmos.

I wrote this book to encourage urban and suburban dwellers to build life experiences by traveling to less light-polluted locations to be awed at the sight of the Milky Way and a star-filled night sky.

I wrote this book to share the happiness of deep-sky observing. If using this book allows you to think "Wow!" just once, it will help you get a better understanding of the cosmos and our place in it. Maybe, just maybe, it can help you start a life-long journey to explore the amazing universe we call home.

Kevin Ritschel
Deep Sky Ranch, Paso Robles, CA
June 2022

An equatorially mounted 6-inch f/8 refractor with the wide-angle 80mm scope mounted piggyback. Most commonly used to survey brighter objects, this is a light and compact grab-and-go telescope.

KEVIN RITSCHEL

USING SMALL TELESCOPES TO EXPLORE THE COSMOS

Congratulations, you've gotten a small telescope or are thinking about getting one because you want to explore the cosmos. OK, what are you going to look at? This book suggests a series of possibilities.

I suggest a list of objects to look for with a small telescope during every season of the year. By small, I mean that I observed every object in this book with an 80mm-diameter refractor telescope, and many of these I also observed in a 62mm-diameter refractor. I'll offer tips below and in the content of the book on how to create and use a "tricked-out" small telescope.

Will these seasonal lists work for a bigger telescope? Absolutely. A bigger telescope usually will show the object in better detail and offer a brighter image. This book suggests some of the best objects to view from the Northern Hemisphere for new observers with any telescope.

Use this book as a planning aid for the types of objects you want to find in the sky. A good star atlas such as *Astronomy* magazine's *The Complete Star Atlas* or its online Star Dome Interactive Star Charts (found at https://astronomy.com/observing/stardome) will provide you with the maps you'll need to find these objects by star-hopping across the sky.

The following are some general rules to follow when trying to use or select a small telescope to observe the sky.

THE 10 COMMANDMENTS OF OBSERVING

COMMANDMENT 1: **The most important thing to succeed — go to a dark-sky site.**

Even huge telescopes won't show you astronomical targets if the night sky you're under is so light-polluted that the sky is brighter than the target. All cities and most suburbs are so light-polluted that most only show a few bright stars and planets. (The joke about the sky in the Los Angeles basin is that the Sun is the only deep-sky object you can see.) The solution is to get out of the city. At a dark-sky site, where there is minimal light pollution, you easily will see the Milky Way with the unaided eye (except during spring when the Milky Way lies below the horizon).

A corollary to the first rule could be to seek out the highest-elevation-site possible. The higher you are, the more of the polluted, dusty, and light-scattering atmosphere is below your observing site and the better the views you'll get. Astronomers have known this since the early 20th century; that's why they build their observatories on mountains.

COMMANDMENT 2: **Get dark adapted before you use a telescope.**

Let your eyes adapt to the dark. This usually takes about five to 10 minutes. During the time you are waiting, *do not* look at a phone or computer screen (unless they can display a faint red image); nor should you use a regular flashlight. Instead, get a red flashlight to find things and read a star chart.

COMMANDMENT 3: **You need a map.**

You can't find your way around a new city without a map; this is even more true in the sky. Maps of the sky are called star charts and it's essential you have either a physical one such as *The Complete Star Atlas* or an online star chart such as https://astronomy.com/observing/stardome.

COMMANDMENT 4: **Don't look for objects that are close to the horizon.**

Look for astronomical objects when they are as close to overhead as possible. This means you are looking through less of Earth's turbulent atmosphere and thus objects will appear brighter and sharper.

COMMANDMENT 5 (COROLLARY TO RULE 4):

Try to observe objects when they are close to the meridian, the imaginary line that runs from north to south through the point directly overhead.

COMMANDMENT 6: **Bring binoculars.**

I have 8x42 or 10x50 binoculars packed in my observing kit and tripod-mounted 10x70s at my home observing site. Several objects in this book, mostly some of the large emission nebulae of summer and winter, show up best through binoculars. This is especially true if you do not have a dedicated wide-field telescope.

COMMANDMENT 7: **You can never over-dress.**

You may think some ski gear is adequate for a long winter night of observing—it is not. Even in California, I take insulated boots; layers of down vests, trousers, and jackets; thermal or fleece under-layers; thinner gloves and thicker mitts; and insulated caps. Hand warmers are a luxury in really cold weather.

COMMANDMENT 8: **After cold-weather clothing, one of the best observing tools you'll ever buy is a reclining lounge chair.**

This is essential if you plan on using binoculars to scan the Milky Way. The lounge chair also doubles as a reasonable bed if you are camping.

COMMANDMENT 9: **Make a checklist of the equipment you need to observe and camp.**

Food and water as well as star charts, eyepieces, filters, binoculars, red flashlights for everyone (especially children), a folding table or tailgate to use as a work surface for your charts and eyepieces, and folding chairs or lounge chairs.

I learned this the hard way. Years ago, I was teaching an observational astronomy class and we packed up our gear to travel to a desert observing site for a New Moon weekend. The class was excited, the sky was pristine, the night wasn't going to be windy or too cold, a fast-food stop before we reached the site assured every belly was properly stuffed. The makings of a perfect night—except, somehow, we forgot to pack the star diagonals for the Schmidt-Cassegrain telescopes we brought.

The wide-angle 80mm refractor rests on a simple alt-azimuth mount. It is optimized for wide-field observations with 2-inch, 82° AFOV eyepieces and a 2-inch diagonal. KEVIN RITSCHEL

COMMANDMENT 10: **You'll see more if you spend a few minutes making an observing list for the night.** This book makes a great start, but after you've had a few sessions under the stars, review what will be visible in the sky during the upcoming night and make a list of your targets in order of right ascension or constellation—you can then be sure to catch objects in the western sky before they set for the night.

TELESCOPE BUYING AND USING RULES

TELESCOPE RULE 1: **Don't buy a telescope based on power. Buy the biggest diameter telescope lens or mirror you can afford or want to move and store.** The power or magnification of a telescope changes when you change the eyepiece. High-power eyepieces have short focal lengths like 4mm or 6mm. The higher the power, the dimmer the object will appear and the smaller the field of view you get. A narrow or small field of view makes it almost impossible to find anything in the sky. Because I wrote this book for observers with small telescopes, everything described can be observed with an 80mm-diameter refractor.

TELESCOPE RULE 2: **Make sure your mount and tripod are steady.** If you can visit a telescope store, go up to the model you're considering and give it a nudge, if it moves all over the place or visibly shakes, try some other models. I also strongly suggest that beginners try before you buy—seek out a local astronomy club and attend one of their star parties to see how telescopes are used and which one or type you like. A telescope should hold steady after you move to a new object and move smoothly as you shift from one object to another.

TELESCOPE RULE 3: **Make sure the telescope can accept 2-inch-diameter eyepieces, or that you can get an adapter to use them.** You will appreciate this feature later as your experience with telescopes and observing grows. Most small quality telescopes have this ability.

TELESCOPE RULE 4: **Try to buy at least one low-power, wide-angle (large apparent field-of-view) eyepiece to make your telescope work better.** These are eyepieces with an apparent field of view (AFOV) of about 68° or higher and a focal length of about 20mm to 40 mm. These aren't cheap, but this will allow you to find objects in the sky much more easily and give far better views of big objects. I strongly suggest that these be 2-inch-diameter eyepieces.

The most common eyepieces I use with my 80mm to 150mm (6-inch) telescopes are 20mm to about 40mm in focal length with AFOVs in the 68° to 82° range. I have also selected telescopes that accept the larger diameter 2-inch standard eyepieces. These are generally standard in higher-end refractors above 3 inches in diameter and reflectors 6 inches or larger.

I should note that if all you want to do is look at the Moon and planets, you don't need the 2-inch eyepieces or the wide AFOV eyepieces. But since you are reading this in a book about observing the deep sky, I assume you want a telescope well suited to look at deep-sky objects.

Also be a little careful buying a low-power eyepiece: don't get one too low in power. Divide the focal length of the eyepiece by the f/ratio of the telescope you want to use. This produces a number called the exit pupil and is the diameter of the beam of light coming into your eye. This has to be below 7mm because that's usually the largest beam of light your eye's dilated pupil can accept. So for my 80mm, f/6 refractor, I could use a 42mm eyepiece as my lowest useful power (42mm/6 equals 7mm).

TELESCOPE RULE 5: **Always use a low-power eyepiece to search for objects in the sky.** You should never attempt to find astronomical objects when using the scope above about 40 to 50 power. The field of view is simply so small that it would be a miracle if you find anything at all. Go ahead and try to find something on the ground during the day using high- and low-power eyepieces to see how much easier low power is to find something.

TELESCOPE RULE 6: **If you want to observe planetary nebulae (dying stars) or emission nebulae (mostly places where stars are being born), you will want to get an OIII filter for your eyepiece.** These filters (pronounced "O-three" filters) thread into the bottom of an eyepiece and greatly help you find and study the glowing gas clouds that are so popular. Some celestial objects are almost invisible without them.

TELESCOPE RULE 7: **Having a red-dot or reflex finder is better than a cheap optical finder scope but having both a decent optical finder scope and a reflex finder is even better.**

WIDE-ANGLE SCOPE RECOMMENDATIONS

Keep it simple for your first telescope. If you or a family member ultimately gets hooked on the hobby, then you can look at buying a bigger scope or getting all the gear needed for imaging.

For specific recommendations, if you are looking for an ultra-portable, wide-angle travel rig such as what I used for this book, a 3- to 5-inch short focus refractor (f/ratio of about 5 or 6) is hard to beat.

If you know you are going to try to find a lot of galaxies, I suggest more aperture to collect more light. For beginners, a 6- or 8-inch Dobsonian is a good all-around scope that will grow with you for years. Yes, a 10-inch, 12-inch, or bigger Dob makes those faint fuzzies much brighter and more detailed, but you have to live with moving something around that's about the size of a traditional water heater. This is not a serious obstacle to those obsessed with this hobby—but your next purchase will likely be a truck or a van to haul all your gear.

A good grab-and-go telescope kit: a 6-inch f/5 reflector with 2-inch eyepieces, a finder scope, and a sturdy, simple mount with slow-motion controls. Not much bigger or heavier than the wide-angle 80mm, this 6-inch telescope gathers 3.5 times more light and uses the same lightweight alt-azimuth mount and tripod. KEVIN RITSCHEL

If you want to image the sky, stick with nightscapes, planets, and piggyback photography with wide-angle and telephoto lenses to start. To image close-ups of the objects discussed in this book is a fairly major financial commitment. A former co-worker assembled a kit to do long-exposure imaging of deep-sky objects through a good 4-inch refractor and produced very impressive, publishable images—and we estimated he spent well more than $25,000 for the camera, telescope, go-to mount, auto-guider, imaging software, and electrically driven focuser, but that story is for another book. Cameras and mounts were getting cheaper before inflation struck, but the expectation is that imaging will get more affordable over time. Technology is making imaging far easier than it was 10 or 20 years ago.

Imaging is a major time commitment as well. A good deep-sky image can mean a minimum of three to six hours of collecting data and even longer in processing that data. Add in the time and expense of traveling to a dark-sky site or to get the narrowband filters and filter wheels to take deep-sky images in a light-polluted area, and each image you produce is the product of a lot of time and expensive gear.

MY FAVORITE TELESCOPES

I have joked that I've never met a telescope I didn't like, and I have owned and used more scopes than I can count, including scopes larger than 30 inches in diameter. All these telescopes provided wonderful views, but the bigger ones were so large I needed a 12-foot ladder to reach the eyepiece, they filled half my garage, and were too heavy for me to transport by myself. I have downsized.

For most observing, I have a sturdy alt-azimuth mount with slow motion controls that I use with either my 80mm f/6 achromatic refractor, a 110mm f/6 apochromatic refractor or a 6-inch f/5 reflector. This grab-and-go telescope kit is perfect if I want to grab a telescope to take a quick look at something or plan a quick visit to a site that's darker than my remote, rural ranch. When I visited Chile, I traveled there with an 80mm refractor.

For more serious observing, I use a 6-inch f/8 apochromatic refractor on a go-to equatorial mount. Why? An 80mm telescope can gather about 130 times what your naked eye can; a 6-inch telescope gathers 3.5 times more than that.

I can co-axially mount any of the small scopes on top of the main 6-inch refractor tube, which allows me to easily compare the view through different-size telescopes. I still have larger, semi-transportable telescopes for when I really want to go deep in the nighttime sky—you never truly get over aperture fever.

Another hint: almost any real telescope will deliver amazing views of dozens or hundreds of objects in the sky, but it takes at least 12 inches of aperture to really get interesting views of objects that start to look like the objects you see in magazines and online (especially if you travel outside our Milky Way to examine distant galaxies past the Virgo Cluster). That said, I really don't want to

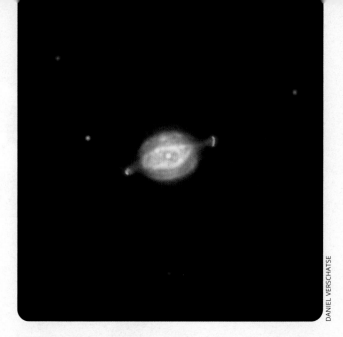

discourage you; you can see a lot even with an 80mm telescope under a dark sky, and a 6- or 8-inch telescope can grow with your observing efforts for many years.

And an important point you should be aware of: The images reproduced in this book and those you see in magazines and online typically reveal far more color and detail than you'll see through a telescope. Color typically comes with long exposures that the human eye simply can't match.

HOW THIS BOOK IS ORGANIZED

I discuss about 160 objects in this book. That is only a fraction of the objects visible in even an 80mm-diameter telescope. The book has four sections, and each has a selection of suggested objects based on the four seasons of the year, starting with winter. Why do we need different sections? As Earth orbits the Sun, the nighttime side of our planet peers off into a different direction of the universe each season, so the stars and the deep-sky objects we see at night differ by season.

Within each section, I list objects from west to east by a coordinate system fixed to the sky similar to longitude and latitude on Earth. West-to-east positions get measured in hours of right ascension (R.A.) from 0 to 24 hours; north-to-south positions get measured in degrees of declination (Dec.) from 90° to –90°. The R.A. and Dec. coordinate grid is fixed on the sky, but since Earth orbits the Sun and faces a slightly different part of the sky every night, different objects with different coordinates appear in the night sky at different times of the year.

Since the astronomical coordinate system is fixed on the sky, I list the coordinates for every object in this book. But if you don't have a go-to telescope or one with setting circles, this information is somewhat superfluous because you can find any object by star-hopping.

LOCATING OBJECTS BY STAR-HOPPING

Star-hopping is easy in theory. You choose the object you want to observe and locate it on a star chart. Then you simply move from star to star on the chart and in parallel in the sky to get to your target's area, point your telescope there, and with a little hunting around, you usually will find the object.

It helps to have a basic knowledge of the sky and constellations, and star-hopping will develop that in you over time. It feels good when you have star-hopped to a target and found it. If you invest the time to learn how to star-hop, the sky will never seem like a stranger again. Working with a basic astronomy class or an astronomy club at a star party can make it much simpler and faster to acquire this skill.

Technology is changing this. Today there are numerous go-to or robotic telescopes that will point your telescope for you—but at a cost. A small, complete, go-to telescope is probably $500 to $1,000 at a minimum and you still need some basic understanding of the sky to align it properly. A robust robotic mount that can hold any of the telescopes needed to view what is in this book is going to be $1,000 to $2,500. The latest advances include mobile-phone-driven night-sky recognition and pointing assist software that are likely the future of observing for beginners.

Go-to telescopes of any type will make your night more productive because you'll be able to slew from object to object in seconds, but there's a learning curve in setting up the basic system and you will not easily gain an appreciation of what and where things are in the sky. For more background on how the objects in the sky move, go to https://www.astronomy.com/observing/get-to-know-the-night-sky. Astronomy.com also has a bunch of videos on astronomical concepts, objects, and equipment.

WHAT'S IN A NAME?

Early observers discovered nearly every object in this book between 1700 and the early 1900s. Astronomical objects are so common that catalogs sprang up to record their positions and names. The earliest was the Messier Catalog, which today is a list of 109 objects compiled by French astronomer Charles Messier in the late 1700s. Because he observed with crude telescopes and eyepieces compared to today's models, it is a classic list of the best deep-sky objects, because that was all he could see.

Later in the 1700s and 1800s, the Herschel family in the United Kingdom and others conducted a systematic survey of the sky (by slow, manual sweeping) and found about 7,000 objects and listed them in what is now called the New General Catalog (NGC). Revisions of the NGC along with a couple of supplemental Index Catalogs (IC) have cataloged more than 12,000 objects and given NGC or IC numbers for almost everything in this book. Professional astronomers still use these lists. Other lists include more star clusters—Collinder, Melotte, and Trumpler to name a few—emission nebulae have been cataloged by Sharpless and Beverly Lynds, and the great America astronomer E.E. Barnard, who also introduced astrophotography to the science in the late 1800s and early 1900s, identified dark nebulae.

In addition to catalog numbers, all objects in this book are listed by their common names, if they have one. Astronomers seem to love lists.

HOW BRIGHT IS IT?

Astronomers also love to quantify things—astronomy is a science after all. So all stars and most non-stellar objects are given a magnitude ranking to quantify their brightness. Magnitude is an indicator of how bright an object appears in the night sky, but it is a little tricky because bigger magnitudes correspond to dimmer objects.

Back when astronomy was a fledgling science, a star was of first importance or magnitude if it was bright. This convention stuck and astronomers divided the naked-eye stars into values of 1st to 6th magnitude, 1st being brightest and 6th being barely bright enough to see. This was made more exact as astronomers became able to measure star brightnesses accurately, and a range of five magnitudes was defined to represent a factor of 100 times difference in brightness.

When astronomers took their new definitions and measurement capabilities to match the real sky, they found that one to six didn't quite work, so some of the brightest stars have values less than one and some even are negative, including the night sky's brightest star, Sirius, which measures magnitude –1.44.

Objects fainter than the faintest stars you can see with the naked eye therefore have larger magnitudes. For example, the galaxy M87 in the spring constellation Virgo, where astronomers imaged the first black hole, comes in at magnitude 8.6. Most of the objects in this book are brighter than 10th magnitude. The faintest object seen by any existing ground-based telescope is 27.7 (with the 8.2-meter Subaru Telescope in Hawaii); the Hubble Space Telescope has a limiting magnitude of 31 and the James Webb Space Telescope should be able to reach 34th magnitude.

How faint you can see with a telescope depends on many factors, including the type of telescope, the power you are using, and the quality of the night sky. You can find a calculator to estimate how faint you can go with your telescope at https://www.cruxis.com/scope/limitingmagnitude.htm

Let's go see something amazing in the sky.

WINTER SKY

For astronomers, winter in the Northern Hemisphere features chilly days and even colder nights. Observing during this season means you'll need more time to prepare for keeping warm. But the reward is well worth it: This season presents a star-filled sky full of celestial wonders unmatched during the other seasons.

The winter sky section of this book covers objects in the night sky with right ascensions between 3 and 8 hours. This part of the sky centers near the bright constellation Orion the Hunter, which appears above the horizon at 8 P.M. from roughly December to March.

The winter sky holds the night sky's most amazing sight visible from the Northern Hemisphere: the Orion Nebula (M42). But it is also rich in other emission nebulae and star clusters that inhabit the winter Milky Way. It's a wonderful time to observe if you don't let the cold get you down!

If you travel to a dark-sky site at this time of year, you'll need serious winter weather gear to stay comfortable. Think about layers. Even in the coastal mountains of Central California, where I made most of the observations for this book, my "deep-space suit" includes insulated boots, two pairs of thick socks, fleece pants, a wind-proof snowmobile bib, a sweatshirt, a down jacket, two wool caps, and a muffler/scarf—and that's for a region that sees normal winter lows in the range of 20° Fahrenheit. In colder climates, comfort-seeking observers typically employ hand warmers, electrically heated socks and/or entire jumpsuits, and other measures that will make you envious if you come unprepared.

Winter observing will place you in a situation you likely have never experienced—even if you are used to cold winters. Deep-sky observing is a sedentary activity and leaves you much more at the mercy of the elements than winter sports like skiing, skating, or hiking where movement generates heat to help keep you warm.

The winter sky has hundreds of targets within reach of an 80mm refractor; a larger telescope will reveal many more. **My Top Ten list,** in order of their total brightness, is: **1** The Pleiades (M45) in Taurus; **2** The Orion Nebula (M42); **3** The neighboring open star clusters M46 and M47 in Puppis; **4** Barnard's Loop in Orion; **5** M37 in Auriga; **6** The California Nebula (NGC 1499) in Perseus; **7** The Monkey Head Nebula (NGC 2174/5) in Orion; **8** The Rosette Nebula (NGC 2237–9, 46) in Monoceros; **9** The Crab Nebula (M1) in Taurus; and **10** Thor's Helmet (NGC 2359) in Canis Major.

Under dark transparent skies, most of these objects can be seen with larger binoculars, though Thor's Helmet might be too small. And you may need an OIII filter and a telescope to see the California Nebula and Barnard's Loop. These two emission regions are large, so their light spreads out and leaves them with a low surface brightness that makes them a challenge to see. It helps to get some observing experience and make sure the night is transparent before trying to track them down.

HOW TO USE THIS MAP

This map portrays the sky as seen near 35° north latitude. Located inside the border are the cardinal directions and their intermediate points. To find stars, hold the map overhead and orient it so one of the labels matches the direction you're facing. The stars above the map's horizon now match what's in the sky.

The all-sky map shows how the sky looks at:

1 A.M. December 15
11 P.M. January 15
9 P.M. February 15

STAR COLORS

A star's color depends on its surface temperature.

- The hottest stars shine blue
- Slightly cooler stars appear white
- Intermediate stars (like the Sun) glow yellow
- Lower-temperature stars appear orange
- The coolest stars glow red
- Fainter stars can't excite our eyes' color receptors, so they appear white unless you use optical aid to gather more light

MAP SYMBOLS

- ⊙ Open cluster
- ⊕ Globular cluster
- □ Diffuse nebula
- ✧ Planetary nebula
- ⬯ Galaxy

STAR MAGNITUDES

- ● Sirius
- ● 0.0
- ● 1.0
- ● 2.0
- • 3.0
- · 4.0
- · 5.0

PAGE	OBJECT NAME	CONSTELLATION	TYPE	R.A.	DEC.	MAG.	SIZE
16	IC 342	Camelopardalis	Spiral galaxy	3h47m	68°06'	9.1	21'
17	The Robin's Egg Nebula (NGC 1360)	Fornax	Planetary nebula	3h33m	–25°52'	9.4	11'x7.5'
18-19	NGC 1365	Fornax	Barred spiral galaxy	3h34m	–36°08'	9.5	10'x6'
22-23	The Pleiades (M45)	Taurus	Open cluster	3h47m	24°07'	1.5	110'
26-27	The California Nebula (NGC 1499)	Perseus	Emission nebula	4h01m	36°37'	6.5	145'x40'
28	NGC 1502	Camelopardalis	Open cluster	4h08m	62°20'	6	9.7'
29	NGC 1746	Taurus	Open cluster	5h04m	23°46'	6.1	42'
30	NGC 1807	Taurus	Open cluster	5h11m	16°31'	7.0	17'
30	NGC 1817	Taurus	Open cluster	5h12m	16°41'	7.7	16'
31	NGC 1851	Columba	Globular cluster	5h14m	–40°03'	7.3	11'
31	M79 (NGC 1904)	Lepus	Globular cluster	5h24m	–24°31'	8.0	8.7'
32	The Flaming Star Nebula (IC 405)	Auriga	Emission nebula	5h16m	34°21'	6.0	37'x10'
33	NGC 1907	Auriga	Open cluster	5h28m	35°20'	8.2	7'
33	M38 (NGC 1912)	Auriga	Open cluster	5h29m	35°51'	6.4	21'
33	NGC 1931	Auriga	Emission nebula	5h31m	34°15'	10.1	3'
33	M36 (NGC 1960)	Auriga	Open cluster	5h36m	34°08'	6.0	10'
33	M37 (NGC 2099)	Auriga	Open cluster	5h52m	32°33'	5.6	24'
34-35	The Crab Nebula (M1)	Taurus	Supernova remnant	5h35m	22°01'	8.4	7'x5'
36-37	The Orion Nebula (M42)	Orion	Emission nebula	5h35m	–5°23'	4.0	65'x60'
38	The Flame Nebula (NGC 2024)	Orion	Emission and dark nebulae	5h42m	–1°51'	—	30'x30'
39	M78 (NGC 2068)	Orion	Emission nebula	5h47m	0°05'	8.3	8'x6'
40-41	Barnard's Loop (Sharpless 2-276)	Orion	Emission nebula	5h28m	–3°58'	5	10°

PAGE	OBJECT NAME	CONSTELLATION	TYPE	R.A.	DEC.	MAG.	SIZE
42	M35 (NGC 2168)	Gemini	Open cluster	6h09m	24°20'	5.1	28'
42	NGC 2158	Gemini	Open cluster	6h07m	24°06'	8.6	5'
43	The Monkey Head Nebula (NGC 2174/5)	Orion	Emission nebula	6h09m	20°40'	6.8	40'x30'
44	Hubble's Variable Nebula (NGC 2261)	Monoceros	Reflection nebula	6h39m	8°45'	9	2'
44	M41 (NGC 2287)	Canis Major	Open cluster	6h46m	–20°45'	4.5	38'
45	The Rosette Nebula (NGC 2237–9, 46)	Monoceros	Emission nebula	6h32m	4°58'	7.0	80'x60'
46	The Christmas Tree Cluster (NGC 2264)	Monoceros	Open cluster	6h41m	9°54'	3.9	20'
47	The Jellyfish Nebula (IC 443)	Gemini	Supernova remnant	6h17m	22°32'	8.8	50'x40'
48	The Seagull Nebula (IC 2177)	Monoceros	Emission Nebula	7h05m	–10°28'	6.2	120'x40'
48	Thor's Helmet (NGC 2359)	Canis Major	Emission nebula	7h19m	–13°14'	—	8'
49	Caroline's Cluster (NGC 2360)	Canis Major	Open cluster	7h18m	–15°38'	7.2	13'
49	NGC 2362	Canis Major	Open cluster	7h19m	–24°57'	3.8	6'
50	M47 (NGC 2422)	Puppis	Open cluster	7h37m	–14°29'	4.4	30'
50	M46 (NGC 2437)	Puppis	Open cluster	7h42m	–14°49'	6.1	23'
50	NGC 2438	Puppis	Planetary nebula	7h42m	–14°44'	10.1	1.1'
51	NGC 2453	Puppis	Open cluster	7h48m	–27°12'	8.3	5'
51	The Skull and Crossbones Nebula	Puppis	Emission nebula	7h52m	–26°27'	7.1	16'

Spiral galaxy IC 342

Lurking in the Milky Way's outskirts in the far northern constellation Camelopardalis the Giraffe lies the galaxy IC 342. This is the fifth object in the Caldwell Catalog, a list that British amateur astronomer Sir Patrick Caldwell-Moore compiled. His goal was to highlight the 109 best sky objects that he felt Charles Messier should have picked up in his list of 109 celestial wonders. Most of the objects, including IC 342, are easy to view, but some pose a real challenge.

IC 342 makes a roughly right triangle with Alpha (α) and Gamma (γ) Camelopardalis. The galaxy glows at magnitude 9.1, so it shows up easily through an 80mm refractor under a dark sky. It's pretty large—about two-thirds the size of the Full Moon—and appears as a circular glow because it is oriented nearly face-on to our line of sight. The galaxy resides among a picturesque foreground of Milky Way stars. Can you see the condensation toward the galaxy's center that marks its nucleus?

IC 342 lies 6.2° west-northwest of open star cluster NGC 1502, which we will visit in 12 pages. If you have a wide-field telescope with a 3° field of view, you can hop from NGC 1502 to IC 342 by traveling two fields due south and 1.3 fields east. As you travel telescopically from the galaxy to the cluster, you may notice a line of stars called Kemble's Cascade, a chance alignment of Milky Way stars. Navigating from object to object as I just described—called star-hopping—is a primary means of finding objects dimmer than naked-eye stars.

Gas and dust in the Milky Way seriously dim IC 342. Astronomers suspect that it would be visible to the naked eye from a dark site if all that Milky Way material wasn't in the way. As it is, you should be able to pick out IC 342 with 10x50 binoculars. I recommend that you mount them on a tripod. This technique proves highly effective at picking out faint fuzzies like galaxies.

IC 342 is a spiral galaxy in the constellation Camelopardalis. Observers sometimes refer to it as the Hidden Galaxy because dust clouds in the Milky Way strongly dim it. T.A. RECTOR/
UNIVERSITY OF ALASKA ANCHORAGE, H. SCHWEIKER/WIYN AND NOIRLAB/NSF/AURA

IC 342
Also known as:
Caldwell 5
Constellation:
Camelopardalis the Giraffe
Right ascension:
3h47m
Declination: 68°06'
Magnitude: 9.1
Apparent size: 21'
Distance: 10.7 million light-years

NGC 1360 is an old star in the process of dying. This stage of stellar evolution produces a planetary nebula as the star puffs off its outer layers and creates a shell-like structure.
ADAM BLOCK

The Robin's Egg Nebula

Moving down to the southern constellation

Fornax the Furnace brings us to the second object in our winter collection: The Robin's Egg Nebula (NGC 1360). Because the nebula lies at a declination of –26°, you won't be able to see it if you live north of 64° north latitude, and realistically you'll have a hard time seeing it unless you observe from latitudes 10° to 15° farther south.

NGC 1360 is a planetary nebula, an object where an old star has cast off its outer atmosphere and set it aglow with ultraviolet radiation from the hot core of the remaining star. So, what you are seeing in the Robin's Egg is a dying star! Spoiler alert—the Sun will go through this evolutionary stage in about 5 billion years and swallow Earth.

To find the nebula, locate the 4th-magnitude star Tau^5 (τ^5) Eridani in the constellation Eridanus and then head 4.2° due south. NGC 1360 appears as a dot, barely larger than a star, through a wide-angle 80mm refractor at low power. You'll need to boost the power to see the planetary's oval-shaped disk. With a big enough telescope, you should see a distinct blue-green tint, which is how the Robin's Egg got its common name. I can see a hint of color through a 6-inch scope. The ability to see color—and what color you see—varies widely among people. The differences in color people report when they look at double stars or deep-sky objects never fails to amaze me. Because this nebula appears so small in an 80mm wide-field instrument, I consider it a moderately challenging object.

If you have a 6-inch or larger instrument, you can find about a half dozen 10th-magnitude and fainter galaxies lying within a 5° circle surrounding NGC 1360. I won't typically list or describe in detail such background objects in this book because seeing anything fainter than 10th magnitude is pretty hard to do with an 80mm telescope.

The Robin's Egg Nebula
Also known as: NGC 1360
Constellation: Fornax the Furnace
Right ascension: 3h33m
Declination: –25°52'
Magnitude: 9.4
Apparent size: 11'x7.5'
Distance: 1,000 light-years

Barred spiral galaxy NGC 1365

Just over 10° due south of NGC 1360 lies a magnificent face-on spiral galaxy. NGC 1365 resides about 56 million light-years from Earth, so the light you see when observing this galaxy left just a few million years after the dinosaurs kicked the bucket. In deep photos, NGC 1365 displays a central bar emanating from its nucleus with a spiral arm starting from each end of the bar. This is a classic example of a barred spiral galaxy.

In an 80mm refractor, this magnitude 9.5 object is visible but a challenge. My small refractor shows a circular haze with a condensation at the nucleus. With averted vision, I can make out ever-so-slightly brighter areas where the bar and spiral arms reside, but I can't discern them well enough to claim that I've seen the arms in this scope. You need at least a 12-inch telescope to see the spiral arms clearly with averted vision, and probably a 16-inch or bigger scope to view the spiral arms. Binoculars show the galaxy as a small hazy spot.

NGC 1365 is easy to find. Locate the chain of 6th-magnitrude stars comprising Chi[1] (χ^1), Chi[2] (χ^2), and Chi[3] (χ^3) Fornacis. The galaxy stands about 1° east-southeast of them. One difficulty in seeing the galaxy well is that it lies fairly far south and climbs only about 20° above the southern horizon from mid-northern latitudes.

NGC 1365

Constellation: Fornax the Furnace
Right ascension: 3h34m
Declination: –36°08'
Magnitude: 9.5
Apparent size: 10'x6'
Distance: 56 million light-years

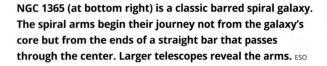

NGC 1365 (at bottom right) is a classic barred spiral galaxy. The spiral arms begin their journey not from the galaxy's core but from the ends of a straight bar that passes through the center. Larger telescopes reveal the arms. ESO

OBSERVING GALAXIES

Seeing galaxies is hard. Although thousands lie within the visual range of amateur telescopes, only a handful are large and bright. Those in the Local Group, such as the Andromeda Galaxy (M31), stand out because they reside only a couple million light-years away. A few galaxies 10 million to 15 million light-years away, including the Whirlpool Galaxy (M51), Bode's Galaxy (M81), and the Cigar Galaxy (M82), appear smaller but still intriguing through small scopes. After that, galaxies get smaller with distance and eventually become mostly featureless glows.

To really see these faint fuzzies, you'll need a telescope with more aperture, which is why you see a lot of 12-inch and larger scopes at star parties. You also can catch a lot of galaxies in more detail if you get into imaging, but that is beyond the scope of this book (forgive the pun), and a major investment to boot. As a ballpark estimate, an imaging kit is going to cost a minimum of $5,000.

With a small scope, it's best to search for a galaxy with low power (20mm focal length or longer). Once you find the target, see if higher powers bring in more detail, though keep in mind that eyepiece focal lengths below about 10mm usually won't deliver more overall detail.

An emission nebula or planetary nebula will show up much better in the eyepiece with a light-pollution filter such as an OIII or Ultra High Contrast (UHC) filter, but they do little to improve contrast on galaxies. So to see galaxies better, you need to buy a bigger scope or go to a star party and ask for a view through someone else's "light bucket." (Most people are happy to oblige.) Still, the views through smaller scopes can be fascinating, and a 4-inch scope can still show you hundreds of galaxies. Be patient. Eventually, you'll learn enough to decide what avenue you want to take in astronomy—go deeper into space with a bigger scope, or off into the land of imaging.

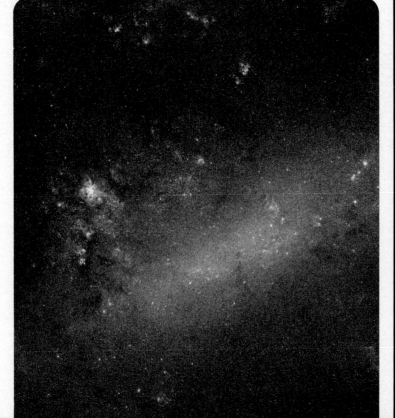

The Large Magellanic Cloud is a satellite galaxy of the Milky Way that straddles the border between the southern constellations Dorado and Mensa.

ESO

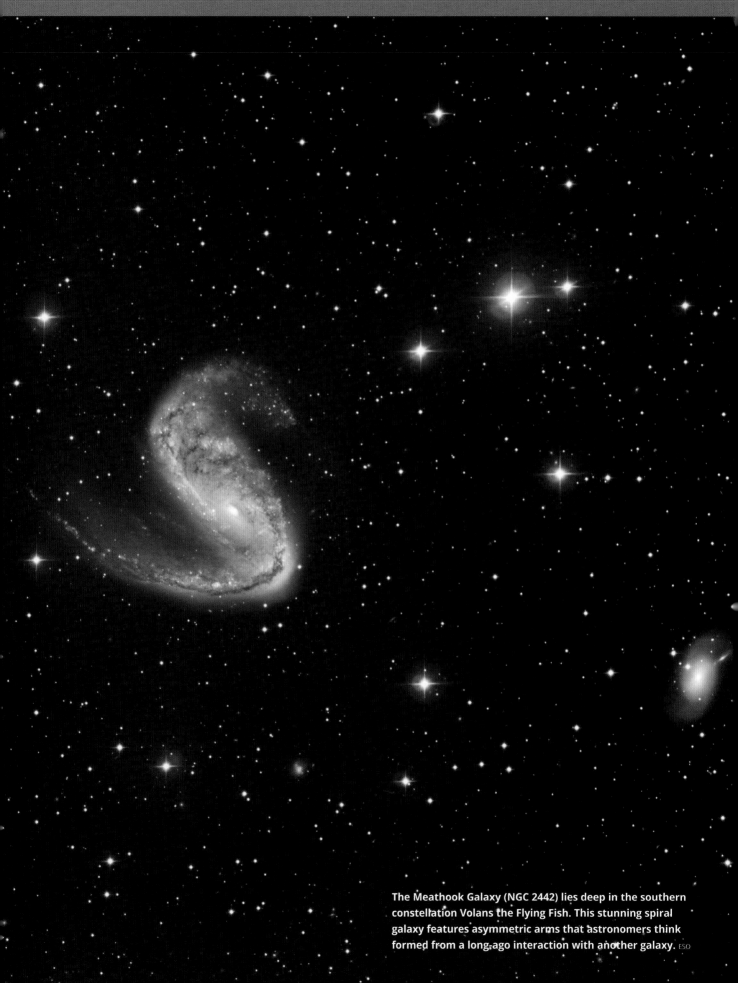

The Meathook Galaxy (NGC 2442) lies deep in the southern constellation Volans the Flying Fish. This stunning spiral galaxy features asymmetric arms that astronomers think formed from a long-ago interaction with another galaxy. ESO

The Pleiades

Sweep north to the constellation Taurus the Bull below the eastern leg of Perseus and you can't miss the Pleiades. This open star cluster, also known as the Seven Sisters and M45, shines so brightly that it has been known since antiquity. You can see the Pleiades with the naked eye even under moderately light-polluted skies. If you need more help, look about one-third of the way from Aldebaran in Taurus to Hamal, the brightest star in Aries. The Seven Sisters has the shape of a little dipper, though it is not the Little Dipper, an asterism in the far northern constellation Ursa Minor. Without a doubt, the Pleiades is one of the most spectacular sights in the night sky.

The cluster consists of hot young stars born less than 100 million years ago. M45 lies some 445 light-years from Earth, so it formed when dinosaurs were romping across our planet. Were the majestic beasts treated to a large emission nebula (a place where stars are born) in the sky close to Earth? Could their lizard brains use such a spectacle as a sign of the changing seasons or for navigation?

This is the type of object where small telescopes have an advantage. When viewed through binoculars or a low-power scope such as my wide-field 80mm instrument with a 22mm eyepiece, the Pleiades looks absolutely spectacular. A close examination will show it holds many double stars, and from a dark-sky site when the Moon is down, several stars in the dipper's bowl seem to have fuzzy edges. This is part of a reflection nebula—starlight reflecting off dust particles in a cloud that happens to be passing through the cluster. Some observers report that they can see the reflection nebula around Merope, the bright star at the bottom of the dipper's bowl, using just 10x50 binoculars.

The cluster's large size plays to a small scope's strengths. M45 is 1.8° across—more than three times the Full Moon's diameter—meaning that something like a wide-field 80mm scope will let you see the whole cluster, not just part of it. Most people find this more satisfying. Even small instruments reveal a bluish tinge to the brightest stars; the color is more obvious if you slightly de-focus the scope.

Be sure to take a look at M45 from the suburbs, Most people see six stars with their unaided eyes. Under a really dark sky, some observers have counted 18. How many can you make out?

The Pleiades (M45) ranks as the third-brightest star cluster in the sky—only the neighboring Hyades and Alpha Persei clusters outshine the beautiful Seven Sisters. TONY HALLAS

The Pleiades
Also known as: The Seven Sisters, M45
Constellation: Taurus the Bull
Right ascension: 3h47m
Declination: 24°07'
Magnitude: 1.5
Apparent size: 110'
Distance: 445 light-years

OBSERVING STAR CLUSTERS

Other than galaxies, open star clusters are the most common deep-sky object, and they are fun to track down. Sometimes they are so big, with the Pleiades being the prime example, that you need a wide-angle scope to see it all. Others lie so far away that they appear tiny, and you will struggle to resolve it into individual stars. Most of the open star clusters visible through small scopes reside in the Milky Way Galaxy, but observatory instruments can resolve open clusters in galaxies beyond our own.

Open clusters can be "open" or "loose," with a lot of space between the stars, or they may be really "tight" with the stars packed close together. Open clusters take many different forms, and the Milky Way holds hundreds of them—all you need to do is seek them out. Another class of star clusters, called globular clusters, are dense balls of many tens of thousands of stars. We'll tackle several of them in the summer sky, where they tend to congregate because they typically lie in the direction of our galaxy's center. Although even big telescopes can't separate stars in a globular's core, small scopes can resolve the edges in some of the closer clusters.

Any telescope will reveal plenty of open and globular clusters. For small, distant open clusters and globular star clusters, bigger telescopes show fainter stars and deliver better resolution.

47 Tucanae is the sky's second-best globular cluster, trailing only Omega Centauri. ESO/M.-R. CIONI/VISTA MAGELLANIC CLOUD SURVEY

The Jewel Box Cluster (NGC 4755) in Crux ranks among the sky's most colorful open clusters. ESO

The California Nebula

Starting at the Pleiades, sweep north into the eastern leg of Perseus the Hero and you'll see three stars forming a slightly curved line. Look about one-quarter of the way from the middle star, 4th-magnitude Xi (ξ) Persei, to the northernmost star, 3rd-magnitude Epsilon (ε) Persei, and you'll find the delicate emission nebula NGC 1499. Dubbed the California Nebula because of its stately shape, it was once considered beyond the reach of amateur astronomers. But that was before better optics and filters made NGC 1499 a pretty easy target.

First try to see the nebulosity with binoculars, then switch to the lowest-power/widest-angle view that you can configure your telescope. On my wide-field 80mm scope, I use a 2-inch diagonal and 2-inch eyepiece of 35mm to 32mm focal length. You should be able to pick out the nebula as a faint, ghostly ribbon stretching up to 2.5° east to west. This works only under a transparent and moonless sky, and the nebula won't have high contrast. You may even need averted vision to see it.

To get a better view, use a light-pollution filter. An OIII works, but an H-beta produces better results. This is one of the few objects that responds well to an H-beta filter. So, if you can see the California with an OIII, it may not be worth the money to buy an H-beta filter even though it helps for this nebula. With a light-pollution filter, the contrast between the glowing emission nebula and the non-glowing sky surrounding it should be high enough for you to see the nebula with direct vision. Note again, however, that this will work only on nights of high transparency and low humidity.

If you orient your telescope so you look along the east-to-west axis, you should see the outline of the state of California. It really does have a strong resemblance.

The California Nebula is an emission nebula, a glowing cloud of hydrogen and helium gases plus traces of other heavy elements, or "metals." Emission nebulae indicate that something is exciting the gas atoms lying between the stars, usually a hot young star (or stars) radiating ultraviolet light that causes the gases to fluoresce like they do in a fluorescent tube or neon sign. In NGC 1499's case, the bright guide star Xi Persei provides this radiation. The California Nebula—the only celestial object named for one of America's 50 states—lies about 1,000 light-years from Earth.

The California Nebula (NGC 1499) is large and faint. Stretching across 2.5°, it can be glimpsed with binoculars under a dark sky but shows up best through a wide-field telescope with an H-beta filter. JASON WARE

The California Nebula
Also known as: NGC 1499
Constellation: Perseus the Hero
Right ascension: 4h01m
Declination: 36°37'
Magnitude: 6.5
Apparent size: 145'x40'
Distance: 1,000 light-years

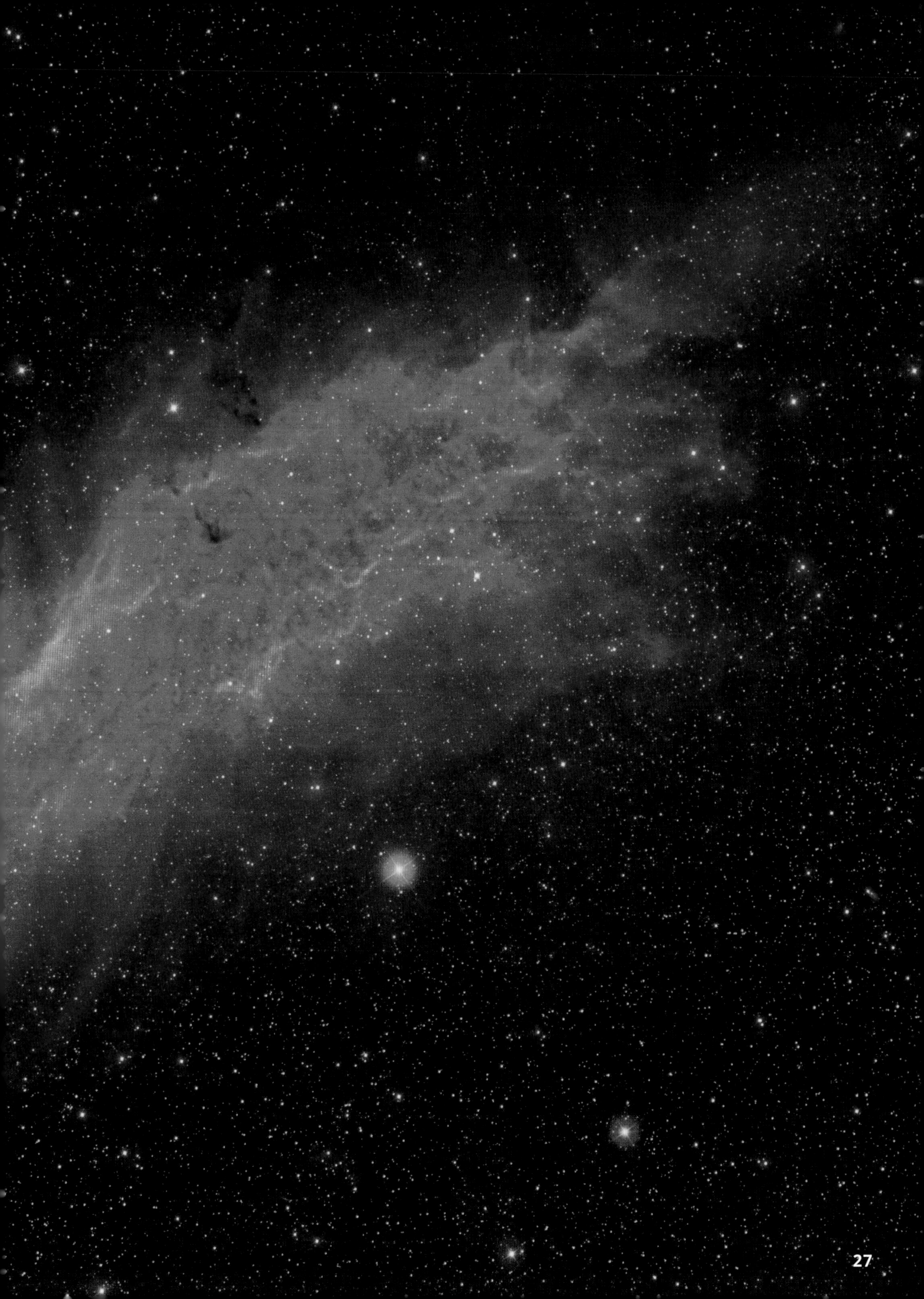

Kemble's Cascade

Technically speaking, NGC 1502 is a young, loosely packed open star cluster holding about 60 stars in the constellation Camelopardalis the Giraffe. It's an easy object through binoculars. But the star of this region has to be Kemble's Cascade, an obvious chain of stars that runs from the cluster off to the northwest. NGC 1502 lies about 3,500 light-years away and holds a number of variable and peculiar stars. A lot of interstellar dust dims it considerably.

The cluster forms an almost equilateral triangle with Alpha (α) and Beta (β) Camelopardalis; NGC 1502 stands at the western apex of this geometric figure. In a small telescope, NGC 1502 resolves into a fairly rich and bright cluster spanning about one-third the size of the Full Moon and set in a rich star field of background Milky Way stars. Although Kemble's Cascade appears to flow into or out of the cluster, this is just a chance alignment typical of asterisms.

By the way, the son-in-law of Johannes Kepler, Jakob Bartsch, popularized the constellation Camelopardalis when he included it in a book he published in 1624.

NGC 1502 is loose open star cluster about 5 million years old. It marks the southeastern edge of the chain of stars dubbed Kemble's Cascade. STEVE COE

NGC 1502

Part of Kemble's Cascade

Constellation: Camelopardalis the Giraffe
Right ascension: 4h08m
Declination: 62°20'
Magnitude: 6
Apparent size: 9.7'
Distance: 3,500 light-years

Open cluster NGC 1746

Moving back to Taurus the Bull, we next target NGC 1746. This scattered ensemble of about 75 stars spans a diameter 40 percent larger than the Full Moon. Look for it between the horns of the Bull, about 5° east and slightly north of 4th-magnitude Tau (τ) Tauri, one of the brighter stars in the northern horn. You can find Tau roughly 7° northeast of the Hyades Cluster, the bright, V-shaped group that forms the Bull's face. (Aldebaran marks the Bull's eye, though it is a foreground object.)

Many people consider NGC 1746 to be a large and loose open star cluster, but in reality, it is an asterism. It holds two distinct associations—NGC 1750 and NGC 1758—that lie within its borders. This wide-field object resides approximately 2,500 light-years from Earth.

NGC 1746 looks like an open star cluster, but it is a chance grouping of stars called an "asterism." Constellation shapes are also asterisms. CC BY-SA 3.0

NGC 1746

Constellation: Taurus the Bull

Right ascension: 5h04m

Declination: 23°46'

Magnitude: 6.1

Apparent size: 42'

Distance: 2,500 light-years

Star clusters NGC 1807 and NGC 1817

Below the horns of Taurus the Bull, between 8° and 9° due east of the constellation's brightest star, 1st-magnitude Aldebaran, sits a pair of open star clusters. NGC 1807 and NGC 1817 make ideal targets for binoculars and small wide-angle telescopes because they reside side by side in the same field.

Shining at magnitudes 7.0 (NGC 1807) and 7.7 (NGC 1817), these clusters show up easily through binoculars. NGC 1807 appears slightly brighter than NCC 1817 even

NGC 1817 and NGC 1807 are two open clusters in Taurus separated by less than the diameter of the Full Moon.

BERNHARD HUBL

though it contains fewer stars. NGC 1807 also appears less concentrated than its neighbor and is a bit harder to pick out from the background stars of the winter Milky Way.

Roughly 10° west of the NGC 1807-NGC 1817 pair lies the V-shaped Hyades star cluster (Caldwell 41). Although Aldebaran appears to be a cluster member marking the southern tip of the "V," it

actually lies only some 65 light-years from Earth compared with the cluster's distance of 150 light-years. The bulk of the Hyades lies west of Aldebaran and is worth a look through binoculars. (The cluster spans some 5.5° and is too big for most wide-field telescopes.) You can start to detect the orange color of the cluster's brighter stars even with 10x50 binoculars.

NGC 1807

Constellation: Taurus the Bull
Right ascension: 5h11m
Declination: 16°31'
Magnitude: 7.0
Apparent size: 17'
Distance: 6,400 light-years

NGC 1817

Constellation: Taurus the Bull
Right ascension: 5h12m
Declination: 16°41'
Magnitude: 7.7
Apparent size: 16'
Distance: 6,400 light-years

Globular clusters NGC 1851 and M79

NGC 1851 lies well south of Orion in the constellation Columba the Dove. At a declination of –40°, this globular cluster resides too far south for observers in the northern United States and Canada. As a Caldwell object, however, it glows quite brightly and makes a fine target for those who live farther south. At magnitude 7.3, it's bright enough to pick up in binoculars and easy through an 80mm scope.

Caldwell 73 spans 11' and appears highly concentrated toward its center. Resolving this cluster—seeing individual stars with black sky between them—proves tough through small scopes when it hangs low in the sky, but it does resolve well in larger instruments.

Some 16° to the north, in the constellation Lepus the Hare, resides the globular M79. It resolves nicely in larger telescopes and is "granular" in my 80mm scope at high power. At magnitude 8.0, M79 shows up well in binoculars.

Globular cluster NGC 1851 resides in Columba the Dove, a southern constellation that nevertheless climbs high enough to view from much of the Northern Hemisphere. NASA/ESA/G. PIOTTO (UNIVERSITÀ DEGLI STUDI DI PADOVA)

Globular cluster M79 lies in Lepus the Hare and stands fairly high on winter evenings. A fairly typical globular, it contains roughly 150,000 stars. NASA/ESA

NGC 1851
Also known as: Caldwell 73
Constellation: Columba the Dove
Right ascension: 5h14m
Declination: –40°03'
Magnitude: 7.3
Apparent size: 11'
Distance: 39,000 light-years

M79
Also known as: NGC 1904
Constellation: Lepus the Hare
Right ascension: 5h24m
Declination: –24°31'
Magnitude: 8.0
Apparent size: 8.7'
Distance: 42,000 light-years

The Flaming Star Nebula

**The Flaming Star
Nebula** (IC 405) in Auriga
the Charioteer may be a
fabulous target for imagers
(as you can see from the
photo above), but it proves
challenging for observers.
To see the nebula visually
requires a dark transparent
sky, and you should hunt
for it through a small telescope only when it stands
near the meridian. I think
including this object in the
Caldwell Catalog was a
stretch. The catalog holds
109 objects that "Messier
should have caught," but
the gifted French observer
can be forgiven for not
spotting the Flaming Star.
Its IC designation gives a
clue that it is harder to
see—the Index Catalog (IC)
is a supplement to the

original NGC of mostly
fainter objects.

It took me multiple
attempts to see IC 405 with
my 80mm telescope. There
is a lesson in persistence
here—if you don't find an
object the first time, try
again on a night with better
transparency and the object
higher in the sky. Of
course, you always will get
a better view when the
Moon is below the horizon.
I made all the observations
for this book when the
Moon was "down."

Once you see the
Flaming Star, you'll agree
that it's a rewarding sight.
The faint, delicate, lacy
nebulosity surrounds the
variable star AE Aurigae,
starting north of this blazing hot star and extending

at least a degree to the
south. Interestingly, this
star formed in the Orion
Nebula (see page 36) more
than 2 million years ago
and is only now passing
through IC 405. The entire
region of glowing gas spans
about 1.5°, three times the
size of the Full Moon. Still,
I don't want to get your
hopes up: even experienced
observers with large ama-
teur telescopes describe the
nebula as extremely faint.

I have seen the nebula
without an OIII or UHC
light-pollution filter under
exceptionally clear conditions, but not as easily as
I did when using such
a filter.

**The Flaming Star
Nebula**

Also known as: IC 405,
Caldwell 31

Constellation: Auriga
the Charioteer

Right ascension: 5h16m

Declination: 34°21'

Magnitude: 6.0

Apparent size: 37'x10'

Distance: 1,500
light-years

Open clusters and more in Auriga

M37 is the brightest of Auriga's clusters. This rich open star cluster holds a large number of stars in a small volume. NOIRLAB/NSF/AURA

A classic open star cluster, M36 features hundreds of stars packed into a relatively small volume. NOIRLAB/NSF/AURA

Slightly fainter than Auriga's other two Messier clusters, M38 still presents a visual treat. NOIRLAB/NSF/AURA

In Auriga the Charioteer, a miraculous stretch of sky holds some of the best open star clusters on Messier's list. Clumped toward the center of the constellation, most of these objects reside within a single wide-field view. The northern portion of the winter Milky Way runs through Auriga, presenting inquisitive observers with star clusters along with a dark nebula and an emission nebula.

The objects listed below appear in order of increasing right ascension. The first of them lies 10° south of the brilliant star Capella.

Open star cluster NGC 1907: This compact open cluster lies just south of our next target, M38. Being so close in the sky, these two objects highlight how star concentrations can change from cluster to cluster.

Open star cluster M38 (NGC 1912): This brighter open cluster lies north and slightly east of NGC 1907. M38 appears more loosely packed, or "open," than its southern neighbor.

Emission nebula NGC 1931: Some 1.7° south-southeast of M38 lies NGC 1931, an emission nebula where new stars are forming out of a cloud of hydrogen gas. In a small telescope, it appears as a mostly circular, glowing ball. An OIII or UHC filter will help you see it. A large gas cloud, IC 417, stretches north of M38 but is not connected to NGC 1931.

Open star cluster M36 (NGC 1960): Just 1° east of NGC 1931 lies another wonderful and bright open star cluster, M36. Through an 80mm scope, it appears

A small region of star formation in Auriga, NGC 1931, lies about 7,000 light-years from Earth. KPNO/NOIRLAB/NSF/AURA/AL AND ANDY FERAYORNI/ADAM BLOCK

as a fairly large irregular cluster of relatively bright stars. You should be able to see all the above objects in one field of view through binoculars or a small wide-angle telescope, forming a marvelous vista, especially when you consider that many of the stars you see in the eyepiece likely have planets orbiting them.

Open star cluster M37 (NGC 2099): If you travel 3.7° east-southeast of M36, you'll come to M37, perhaps the most appealing open star cluster in this vista. M37 is a rich

concentration of about 150 stars in a symmetrical pattern that spans about 25 light-years. Unlike many other open clusters, M37's stars appear evenly distributed. And its edge stands out well despite its position against a rich Milky Way background. Depending on your scope, you might find M37 on one edge of your field and M36 on the opposite edge, but you probably won't be able to see all the way to M38. Binoculars provide the best way to see the three Messier clusters at the same time.

M37

Also known as: NGC 2099
Constellation: Auriga the Charioteer
Right ascension: 5h52m
Declination: 32°33'
Magnitude: 5.6
Apparent size: 24'
Distance: 4,500 light-years

The Crab Nebula

If you swing back south into Taurus the Bull, you can chase down one of the most significant astrophysical objects in the Milky Way. The Crab Nebula (M1) marks the site where Chinese and Japanese astronomers saw a star die in A.D. 1054. At the time, this "supernova" appeared bright enough to see during daylight.

The Crab Nebula shows up easily through an 80mm telescope under a dark sky. Simply point your telescope 1.1° northwest of magnitude 3.0 Zeta (ζ) Tauri, the star at the tip of the Bull's southern horn. There you will find a fairly small oval ball. The glowing gas you see represents the tattered remains of the star that died nearly a millennium ago (at least in Earth time; the Crab's light takes 6,500 years to reach our planet). The gas expands outward at more than 3 million mph, and the nebula now measures 10 light-years across.

M1 does not shine brightly, so you'll likely need to view it when the Moon is out of the sky. Through a small telescope, M1 appears as a fairly uniform glow. An OIII filter helps; in fact, using a large amateur telescope with such a filter shows the object's crablike legs that images so nicely reveal. Studying the Crab Nebula has helped astronomers understand how massive stars run out of fuel and die, and specifically what such explosions leave behind—neutron stars, pulsars, and shock waves that rip through the interstellar medium and can trigger star formation.

The Crab Nebula

Also known as: M1, NGC 1952

Constellation: Taurus the Bull

Right ascension: 5h35m

Declination: 22°01'

Magnitude: 8.4

Apparent size: 7'x5'

Distance: 6,500 light-years

When stars with masses greater than 12 times that of the Sun run out of nuclear fuel, they collapse and then explode as a supernova. The Crab Nebula (M1) is the remnant of such a supernova seen nearly 1,000 years ago. ESO/MANU MEJIAS
Inset: The Crab's heart explodes with detail. ESA/HUBBLE & NASA

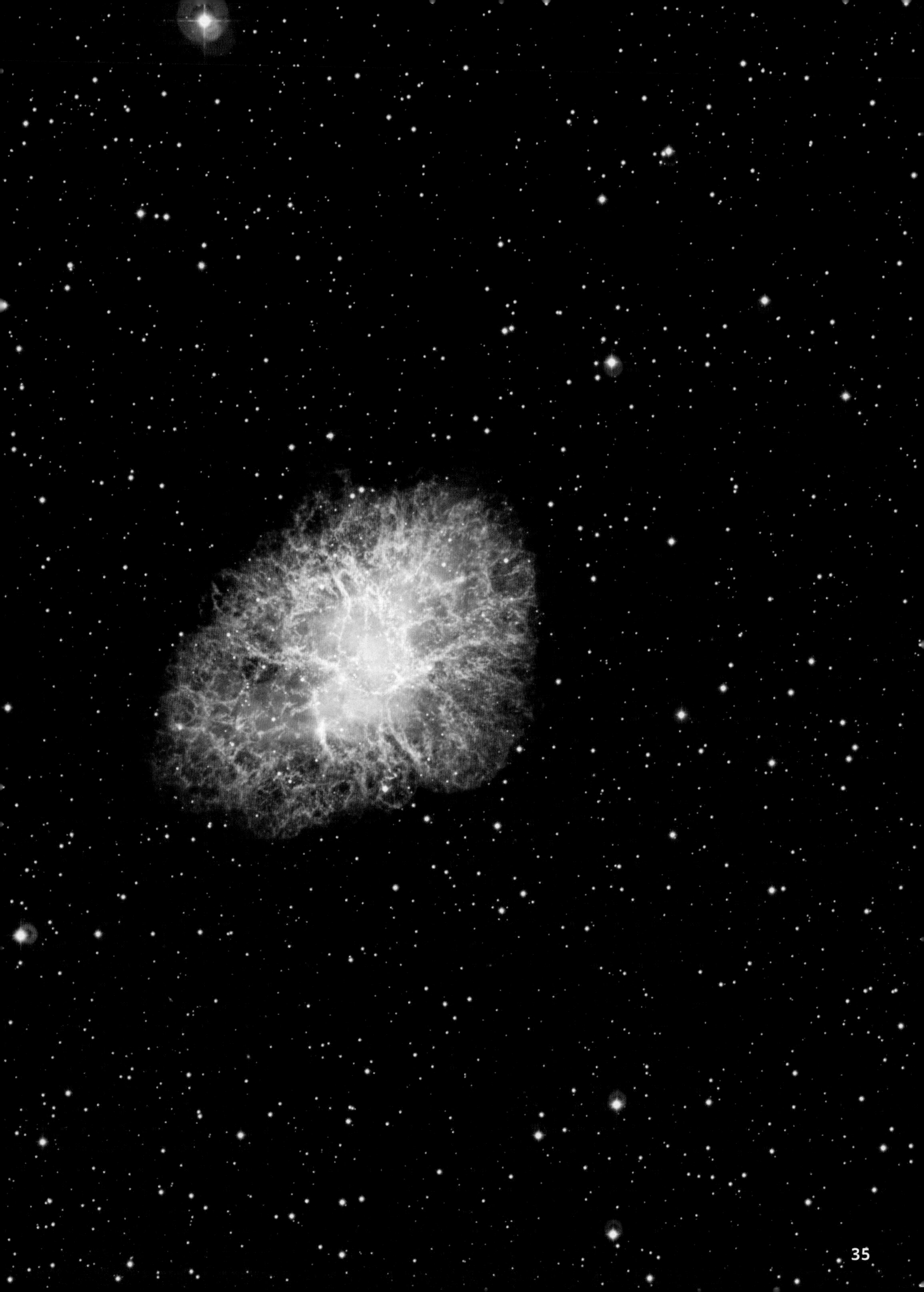

The Orion Nebula

We just examined the Crab Nebula (M1) in Taurus, where a star died in a massive explosion witnessed nearly 1,000 years ago. Now, we head south to the constellation Orion the Hunter, where we can investigate one of the best places to see where stars are born. The Orion Nebula (M42) never fails to impress.

You can see M42 with the naked eye, binoculars, or a telescope. (Eagle-eyed observers can detect the nebula's fuzzy glow even under a moderately light-polluted sky.) Start at Orion's Belt. This distinctive pattern of three closely spaced 2nd-magnitude stars slashes across the Hunter's midsection. Head due south from the belt's middle star and you'll come across Orion's Sword, which appears to the eye as three stars of different magnitudes. Examine the sword's middle star and you'll find it to be fuzzy; binoculars hint at its true nature as a fan-shaped gaseous object, or nebula. Pointing a telescope at M42 reveals a complex cloud of glowing gas and dusty zones with many embedded newborn stars. In the Orion Nebula, gravity compresses massive amounts of gas and dust to forge new stars.

Just about any telescope can deliver a great view of M42. Because the nebula lies only about 1,300 light-years from Earth, it appears bigger, brighter, and more detailed than most astronomical objects. In fact, M42 covers an area six times that of the Full Moon. The larger the telescope you use, the more you can zoom in to individual sections. But most people enjoy viewing M42 at a power low enough so they can see the whole picture and not over-magnify it to explore just a part of the story. A 6-inch f/8 telescope with a 20mm to 35mm eyepiece seems about ideal to me. Smaller scopes like my wide-field 80mm instrument show a respectable amount of detail with a 17mm wide-field eyepiece but also reveal details in other parts of the sword. Reflection nebulae, emission nebulae, and star clusters populate the whole region.

The Orion Nebula is one object that you can view for extended periods without tiring of it. I encourage you to look for at least two features. First, target the Trapezium at M42's bright core. Using moderate power, you'll see the nebula's central region resolve into four bright stars arranged in a small boxlike pattern (a trapezoid). A couple of other stars show up through larger scopes. These are just the tip of the iceberg, however—a whole cluster is buried in there,

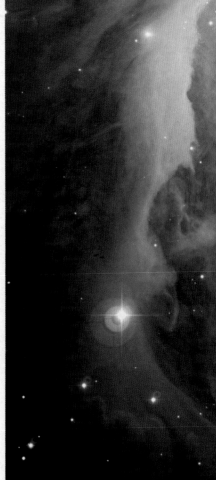

The Orion Nebula

Also known as: M42, NGC 1976

Constellation: Orion the Hunter

Right ascension: 5h35m

Declination: –5°23'

Magnitude: 4.0

Apparent size: 65'x60'

Distance: 1,300 light-years

---

obscured by dust. The second feature to look for is color. In smaller scopes, people often report seeing a greenish tint; in larger instruments, the brighter portions show a tinge of red. What color do you see?

A true marvel, the Orion Nebula (M42) is a massive cloud of gas and dust forming new stars. ESO/IGOR CHEKALIN **Inset: The Trapezium system lies at the nebula's core.** C.R. O'DELL AND S.K. WONG (RICE UNIVERSITY) AND NASA/ESA

The Flame Nebula

The Flame Nebula (NGC 2024) is an emission nebula just east of the easternmost star in Orion's Belt, Alnitak. It's bright enough to see in binoculars but tough to make out against the star's glare. The spectacular Horsehead Nebula juts into the nebulosity. ESO/DSS2

The Flame Nebula

Also known as: NGC 2024

Constellation: Orion the Hunter

Right ascension: 5h42m

Declination: –1°51'

Magnitude: —

Apparent size: 30'x30'

Distance: 1,400 light-years

Head some 3.5° north-northeast of the Orion Nebula and you'll come to another stellar nursery, the Flame Nebula (NGC 2024). The easiest way to get there is to start with the belt's easternmost star, magnitude 1.7 Alnitak (Zeta [ζ] Orionis). Center the star at low power and then increase the magnification to 50x or more. Now move Zeta just outside the field of view to the west. You should see NGC 2024 as a patch of nebulosity immediately east of Zeta. A 6-inch telescope reveals the dust lane that cuts the Flame Nebula in half.

With my wide-field 80mm scope, it's a challenge to see the Flame Nebula unless I move Zeta out of the field. An OIII or UHC filter increases the contrast and allows you to better view the nebula. A larger scope makes it significantly easier to see the nebulosity. Although this object proves challenging through the 80mm instrument, it's quite easy in a 6-inch and will knock you in the eye with a 17.5-inch Dobsonian. The nebula spans 0.5°, making it as large as the Full Moon.

NGC 2024 lies about 1,400 light-years from Earth. It is an emission nebula where gas is condensing into stars. In fact, much of Orion is a hotbed of star formation. A massive cloud of gas and dust known as the Orion Molecular Cloud Complex fills much of the constellation. The Flame Nebula glows because strong ultraviolet radiation from the hot star Alnitak ionizes the cloud's gaseous hydrogen. When the ions recombine, they emit visible light.

Emission nebula M78

Alnitak serves as a useful guide to another bit of nebulosity in Orion. Slide your telescope 2.5° north-northeast of the star and you'll land on M78. With a wide-field 80mm instrument, you merely have to push the scope a single field of view from Alnitak in the 11 o'clock direction. This scope shows the nebula easily even at low power (from a dark sky, of course). But you'll want to boost the magnification to 30x or more to see some detail. Then you'll observe two distinct patches: M78 (NGC 2068) is larger and NGC 2071 smaller.

M78 is a reflection nebula—in fact, it's the brightest one in the sky. Here, dust in the cloud reflects starlight, though emission patches dot the complex as well. With higher powers, the main part of M78 splits again and you can faintly see a central star in each section, like eyes staring at you. Unfortunately, filters do not improve the view of reflected starlight.

One of the brighter sections of Barnard's Loop (see the following spread) lies less than 2° east of M78. The 9th-magnitude open star cluster NGC 2112, which stands 1.8° east and slightly north of the reflection nebula, is embedded in Barnard's Loop.

M78
Also known as: NGC 2068
Constellation: Orion the Hunter
Right ascension: 5h47m
Declination: 0°05'
Magnitude: 8.3
Apparent size: 8'x6'
Distance: 1,400 light-years

M78 (NGC 2068) is a small reflection nebula in Orion where dust reflects starlight in our direction. Once you find M78 in a telescope, you are well on your way to locating Barnard's Loop. ESO/IGOR CHEKALIN

Barnard's Loop

When you look at a wide-field deep image of Orion, like the one on the opposite page, you'll notice a half-circle of red nebulosity on the constellation's eastern (left) side. That dim arc is a supernova remnant—a much larger and older version of the Crab Nebula (see page 34). It has expanded so much that it has pushed parts of the Orion Molecular Cloud Complex toward the east and in the process compressed and excited the gas into emitting light. Astronomers think the nebula is about 2 million years old. If you have a telescope with a clock drive, mount a DSLR on top and take a 5- to 10-minute exposure of the entire constellation. Such a piggyback image, probably the simplest form of deep-sky astrophotography, will show Barnard's Loop quite well. But what is easy to photograph is hard to see visually.

As I described on the previous page, Barnard's Loop lies directly east of the reflection nebula M78, and the open star cluster NGC 2112, which resides nearly 2° east and a bit north of M78, sits in Barnard's Loop. After you find M78, look for the Loop with binoculars. This is often the easiest way to spot the structure.

The great English-German astronomer William Herschel likely saw Barnard's Loop visually. He published drawings of areas where he thought nebulosity existed, and one drawing contained the sky where the Loop lurks. American astronomer Edward Emerson Barnard got naming privileges when he took long-exposure photographs in 1894 that proved the existence of this hard-to-see object.

Credible observers have reported seeing the Loop with their naked eyes, especially if they held an H-beta filter in front of their eyes. Binoculars and small telescopes can show the object, but in these cases, the air must be absolutely transparent, the site as dark as possible, and the Moon below the horizon. The recipe for finding Barnard's Loop with a small portable scope is to use an eyepiece with the lowest power and widest apparent field of view you can lay your hands on. Employ an H-beta filter if you can, but if you don't have one, try an OIII filter instead.

Perhaps the best view of the Loop I ever had was through a 10-inch Dobsonian with an H-beta filter and a 2-inch, 32mm ocular on a cold and exceptionally clear winter's night. I saw the Loop clearly with direct vision. The quality and darkness of the night likely surpasses every other factor determining success in this effort. One thing is certain: If you succeed in observing Barnard's Loop, you have conquered one of the most challenging visual tasks in deep-sky observing.

You also may want to try another trick for finding this faint nebulosity. Once you locate M78 or NGC 2112, move your telescope left to right a couple degrees to each side. As you swing your scope, you may notice the brightness of the background change—you've found Barnard's Loop. Faint objects are often easier to notice when they are moving.

Barnard's Loop

Also known as:
Sharpless 2-276

Constellation: Orion the Hunter

Right ascension: 5h28m

Declination: –3°58'

Magnitude: 5

Apparent size: 10°

Distance: 1,400 light-years

Barnard's Loop represents the leftovers of an exploded star. This supernova remnant got its name from the famous American astronomer E.E. Barnard, who first photographed the huge nebula in 1894. CHRIS SCHUR

Star clusters M35 and NGC 2158

Look northeast of Orion and you'll find the distinctive constellation Gemini the Twins. Among Gemini's deep-sky objects, we only have space to cover a few. But the grand open star cluster M35 is a treasure. This object lies in western Gemini, but it's easier to locate if you start in the eastern part of the constellation with magnitude 1.6 Castor (Alpha [α] Geminorum), the fainter of the "twin" stars Castor and Pollux.

Then follow the line of stars representing the northern twin until you reach 3rd-magnitude Mu (μ) Geminorum. From Mu, locate two fainter stars pointing west toward the horns of Taurus. M35 forms the northern vertex of a roughly equilateral triangle with these stars. Scout the area with binoculars to find M35 and then point your telescope in the cluster's direction.

My wide-field 80mm telescope at 29x resolves the magnitude 5.1 cluster completely. M35 is big, nearly the size of the Full Moon, in part because it resides a modest 3,000 light-years from Earth. If you examine it closely, you should notice a smaller fuzzball immediately to its southwest that appears almost attached. This is NGC 2158, an open cluster that lies some 8,000 light-years beyond M35 and thus looks much smaller and more compact.

NGC 2158 glows at magnitude 8.6 and looks like a milky glow through small scopes, perhaps grainy but not resolved at higher power. You need an 8-inch telescope to see it as a compressed ball of mostly faint stars.

Despite M35's prominence, it doesn't have a common name. Does anything come to your mind? It reminds me of a giant luminescent Volvox, the microscopic colonies of green algae I tormented in high school biology lab.

Two other clusters lie nearby. Just 0.6° due west of NGC 2158 resides magnitude 8.4 IC 2157. It's the same size as its eastern counterpart but has only 5 percent as many stars. The second, NGC 2129, lies 1.1° southwest of IC 2158. Glowing at magnitude 6.7, it appears as a haze of stars with two brighter ones at its core.

M35 (NGC 2168) is a bright and rich open star cluster in Gemini. You can resolve this cluster through a telescope, but not the smaller background cluster NGC 2158 snuggled against its southwestern corner. N.A. SHARP/NOIRLAB/NSF/AURA

M35
Also known as: NGC 2168
Constellation: Gemini the Twins
Right ascension: 6h09m
Declination: 24°20'
Magnitude: 5.1
Apparent size: 28'
Distance: 3,000 light-years

NGC 2158
Constellation: Gemini the Twins
Right ascension: 6h07m
Declination: 24°06'
Magnitude: 8.6
Apparent size: 5'
Distance: 11,000 light-years

A binocular object in a dark sky, the Monkey Head Nebula (NGC 2174/5) in Orion lies 4° south of M35 in Gemini. CARSTEN FRENZL/CC BY-2.0

The Monkey Head Nebula

Slide 3.9° south of M35 and you'll be back in the northern reaches of Orion the Hunter, where a splendid emission nebula and open star cluster, NGC 2174 and NGC 2175, respectively, reside. Once you find M35 in binoculars, go a little less than one field of view to the south. The soft circular glow of NGC 2174 shows up easily, assuming you are looking from under a dark sky, in the Hunter's upraised arm. A wide-field 80mm refractor likewise delivers a nice view.

NGC 2175 appears as a small knot of stars offset slightly below NGC 2174. But visually, the emission nebula encloses the entire cluster, making it look like a single object. Images of the nebula reveal two slightly darker dust clouds that reminded one observer of a pair of eyes. The description grew from there until it encompassed a primate's entire head. The Monkey Head Nebula often gets overlooked because of the other spectacular emission nebulae within Orion, but it is pleasantly bright and well worth a visit. The hot stars embedded in the nebula help sculpt its intricate shape through their prodigious output of high-energy ultraviolet radiation and fierce stellar winds.

The Monkey Head Nebula

Also known as: NGC 2174/5

Constellation: Orion the Hunter

Right ascension: 6h09m

Declination: 20°40'

Magnitude: 6.8

Apparent size: 40'x30'

Distance: 6,400 light-years

Hubble's Variable Nebula and M41

Our next target stands 11° east of 1st-magnitude Betelgeuse in Orion, across the border in neighboring Monoceros the Unicorn. Hubble's Variable Nebula (NGC 2261) may be small, but it's a captivating object that's easy to see in a wide-field 80mm scope.

The nebula glows brightly enough that you can see some structure at higher magnifications. Find it with about 30 power and then boost it to 50x or higher with a larger scope. The dust cloud that forms this comet-shaped nebula reflects light from the variable star R Monocerotis,

which lies at its tip. As the star varies in brightness, the nebula's size varies. Although William Herschel discovered this object in 1783, American astronomer Edwin Hubble observed its variations, which is how it got its common name.

Next, head south to Canis Major the Big Dog and find the brightest star in the night sky, Sirius. With binoculars, drop south 4° (less than the field of view in most binoculars) and you'll come upon an obvious star cluster, M41 (NGC 2287). Next, set aside the binoculars and try to spot the magnitude 4.5

cluster with your naked eye. It should appear as a fuzzy brighter area, particularly if you use your hand to block dazzling Sirius. M41 appears slightly larger than the Full Moon and shows up well through binoculars or a telescope, even in moderately light-polluted skies, despite its distance of 2,300 light-years.

An 80mm telescope resolves M41, showing many of its roughly 100 stars. Astronomers estimate that the cluster will dissipate in about 500 million years. Although the bright orange star near the cluster's center is only 1,500 light-years away, M41 does hold several orange-colored red giant stars.

Hubble's Variable Nebula

Also known as: NGC 2261, Caldwell 46
Constellation: Monoceros the Unicorn
Right ascension: 6h39m
Declination: 8°45'
Magnitude: 9
Apparent size: 2'
Distance: 2,500 light-years

M41

Also known as: NGC 2287
Constellation: Canis Major the Big Dog
Right ascension: 6h46m
Declination: –20°45'
Magnitude: 4.5
Apparent size: 38'
Distance: 2,300 light-years

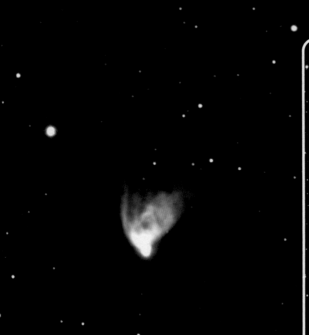

Hubble's Variable Nebula (NGC 2261) got its name from Edwin Hubble, the same astronomer who gave his name to the Hubble Space Telescope. The nebula gets larger and brighter as a nearby variable star brightens. It was the first object astronomers imaged with the 200-inch Hale Telescope. N.A. SHARP, J. MCGAHA/NOIRLAB/NSF/AURA

The large and bright open star cluster M41 (NGC 2287) proves an easy target in binoculars. Can you see it with your naked eye? NOIRLAB/NSF/AURA

The Rosette Nebula

The Rosette Nebula (NGC 2237–9, 46) is a large but faint emission nebula in Monoceros. Binoculars show it under a dark sky. KPNO/NOIRLAB/NSF/AURA/WALTER MULLIGAN/FLYNN HAASE

Monoceros the Unicorn looks inconspicuous to the naked eye, with just three stars brighter than 4th magnitude. But the deep-sky objects it holds more than make up for its lack of prominence. With your binoculars, scan 4° south-southwest of Hubble's Variable Nebula (see the opposite page) or 9° east-southeast of Betelgeuse in Orion. There you should spot the soft glow of a large emission nebula with a fairly conspicuous star cluster at its center. You have just found the Rosette Nebula (NGC 2237–9, 46).

If you now point your telescope to the same location, you'll easily see the star cluster NGC 2244. The larger emission nebula, however, appears far more delicate and can be seen only from a dark site. Carefully look outside the cluster and you should see the emission nebula's outline. An 80mm telescope shows it nicely, but you can punch up the contrast with an OIII filter to make it far more visible. Because the Rosette is so big, spanning 80'x60' (more than twice the Full Moon's diameter), a wide-angle telescope helps. You also should notice that the center appears devoid of nebulosity—the hot stars in the central cluster push the gas outward and provide the ultraviolet energy needed to ionize the gas and make it glow.

The Rosette Nebula is the official state astronomical object of Oklahoma, probably because their official flower, a red rose, looks a lot like deep images of the Rosette.

The Rosette Nebula

Also known as: NGC 2237–9, 46, Caldwell 49

Constellation: Monoceros the Unicorn

Right ascension: 6h32m

Declination: 4°58'

Magnitude: 7.0

Apparent size: 80'x60'

Distance: 5,200 light-years

The Christmas Tree Cluster

Five degrees north-northeast of the Rosette Nebula and a mere 1° north of Hubble's Variable Nebula lies a conspicuous naked-eye star cluster and a weak emission patch visible through binoculars. When oriented with north up, the cluster resembles the stylized outline of an upside-down Christmas tree, hence its name. An 80mm telescope shows the cluster nicely.

A large but faint emission patch surrounds the Christmas Tree Cluster (NGC 2264). An associated dark nebula appears silhouetted against this glow, creating the appearance of a cone just off the cluster's southern end. Although the star cluster proves easy to see, the Cone Nebula glows faintly and shows up only through large telescopes.

The stars of the Christmas Tree Cluster (NGC 2264) stand out at the top of this image, while the Cone Nebula appears as a distinct indentation at the bottom. ESO

The Christmas Tree Cluster

Also known as: NGC 2264

Constellation: Monoceros the Unicorn

Right ascension: 6h41m

Declination: 9°54'

Magnitude: 3.9

Apparent size: 20'

Distance: 2,300 light-years

The Jellyfish Nebula

North of Monoceros and on a line joining Mu (μ) and Eta (η) Geminorum lies a large but faint emission patch, the supernova remnant known as the Jellyfish Nebula (IC 443). This object proves to be a challenge through my wide-field 80mm telescope. With an estimated magnitude of 8.8 and spanning 50'x40', its light spreads out so much that it has a low surface brightness. The nebula's brightest part appears crescent-shaped and measures about 30'x15', or about the size of a First Quarter Moon.

To see this object clearly, you need a wide field, low power, an OIII filter, and an exceptional night. You will probably see it only with averted vision unless you use a big scope. Astronomers estimate the supernova that gave birth to this remnant exploded between 3,000 and 30,000 years ago at a distance of about 5,000 light-years. Does the accompanying photograph look like a Jellyfish to you?

The Jellyfish Nebula

Also known as: IC 443
Constellation: Gemini the Twins
Right ascension: 6h17m
Declination: 22°32'
Magnitude: 8.8
Apparent size: 50'x40'
Distance: 5,000 light-years

The Jellyfish Nebula (IC 443) in Gemini is a supernova remnant, the remains of a massive star that exploded thousands of years ago. BOB AND JANICE FERA

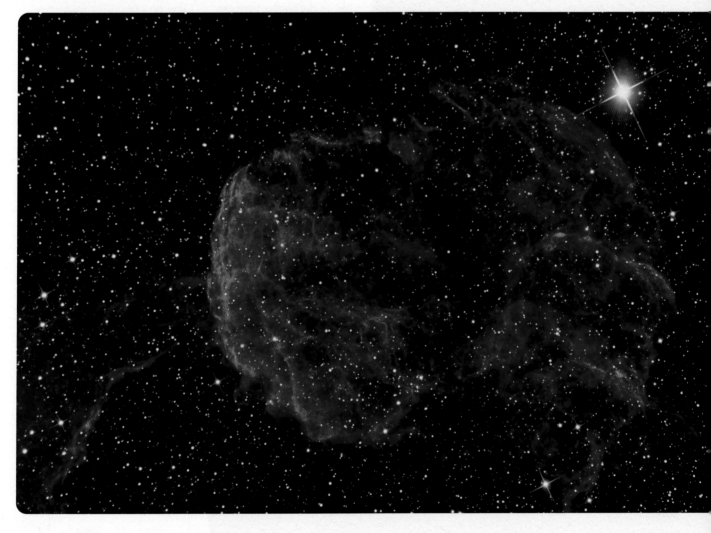

The Seagull Nebula and Thor's Helmet

The Seagull Nebula (IC 2177) lies on the border between Monoceros and Canis Major. Although faint, it shows up in a small telescope. ESO/DSS2

SMALL TELESCOPE. ESO/DSS2

The small emission nebula Thor's Helmet (NGC 2359) lies 4° southeast of the Seagull Nebula. An OIII filter shows off this nebula best. ESO/DSS2

Travel back to Monoceros the Unicorn and head to its southern border with Canis Major the Big Dog. There, just 7° northeast of Sirius, lies the Seagull Nebula (IC 2177). This large emission region has a low surface brightness because its 6th-magnitude total brightness spreads out over an area five times that of the Full Moon. It shows up with direct vision in my wide-field 80mm scope and a 22mm eyepiece. If you can't see it directly, try an OIII filter, which should reveal the long arch of nebulosity that represents the Seagull's wings in flight.

Just 4° southeast of the Seagull lies Thor's Helmet (NGC 2359). This nebulous region shows up through an 80mm refractor, but wait until you see the effect an OIII filter has. With no filter, the nebula is hardly there—thread-in an OIII filter and the view radically improves. The nebulosity takes on a beautiful, intricate shape, and even the horns become visible.

The Seagull Nebula
Also known as: IC 2177
Constellation:
Monoceros the Unicorn
Right ascension:
7h05m
Declination: –10°28'
Magnitude: 6.2
Apparent size: 120'x40'
Distance: 3,800
light-years

Thor's Helmet
Also known as: NGC
2359
Constellation: Canis
Major the Big Dog
Right ascension: 7h19m
Declination: –13°14'
Magnitude: —
Apparent size: 8'
Distance: 12,000
light-years

Caroline's Cluster and NGC 2362

Caroline Herschel discovered Caroline's Cluster (NGC 2360) in 1783. It proves to be an easy object through binoculars.

NASA/ESA/T. VON HIPPEL (EMBRY-RIDDLE AERONAUTICAL UNIVERSITY)

Open clusters often show up nicely through binoculars, and Caroline's Cluster (NGC 2360) is no exception. Named after its discoverer, Caroline Herschel (the younger sister of William Herschel and an accomplished astronomer in her own right), this cluster lies 8° east-northeast of Sirius. NGC 2360 holds about 80 stars and appears symmetrical and well concentrated through binoculars and small scopes.

Scan 9° south of Caroline's Cluster and you'll come to a far brighter open star cluster: NGC 2362. Otherwise known as the Tau Canis Majoris Cluster, its stars surround the magnitude 4.4 blue supergiant star Tau (τ) Canis Majoris. This star lies at the northeastern end of the cross at the southern end of Canis Major.

Thanks to Tau, this cluster ranks as the ninth-brightest open cluster in the sky. The star shines roughly 50,000 times brighter than the Sun, which is why it appears as bright as it does despite lying nearly 5,000 light-years from Earth.

Caroline's Cluster

Also known as: NGC 2360, Caldwell 58
Constellation: Canis Major the Big Dog
Right ascension: 7h18m
Declination: –15°38'
Magnitude: 7.2
Apparent size: 13'
Distance: 3,700 light-years

NGC 2362

Also known as: The Tau Canis Majoris Cluster, Caldwell 64
Constellation: Canis Major the Big Dog
Right ascension: 7h19m
Declination: –24°57'
Magnitude: 3.8
Apparent size: 6'
Distance: 4,800 light-years

The open star cluster NGC 2362 in Canis Major looks beautiful in small scopes. Its brightest star, Tau Canis Majoris, lies at its center and shines at magnitude 4.4. ESO/DSS2

Bright star clusters in Puppis

These three objects fit in a single field of view through a wide-angle telescope, a magnificent vista in the constellation Puppis the Stern worth spending your time on. Use your binoculars and scan some 13°, or about two fields of view, due east of Sirius. Even with your naked eye, you can tell there's something in this vicinity. Let your eyeballs explore and see what fuzzy areas you can find. A small scope shows all three objects well.

The two clusters really stand out in any instrument. The western cluster, M47 (NGC 2422), holds about 30 fairly bright stars that separate easily. The one to the east, M46 (NGC 2437), looks profoundly different. It has about 100 easily visible and evenly distributed stars.

M46 also stands out because it appears to contain a planetary nebula, NGC 2438. It's rather large for a planetary, measuring 1.1' in diameter, but that is still too small for a wide-field telescope. To see NGC 2438, place M46 in your field of view and pump up the power to 70x or more. You'll see a delicate round ball some 7' north of the cluster's center.

Although all three objects lie near each other in the sky, this is a chance

M46 and M47 form an impressive vista in binoculars. In this view, M47 lies at the center and M46 stands near the left edge. A telescope reveals the planetary nebula NGC 2438 in front of M46. ESO/DSS2

alignment. M47 resides 1,600 light-years from Earth while M46 dwells 4,900 light-years distant— three times farther away. The planetary NGC 2438 lies only about 1,400 light-years from Earth, so it's a foreground object.

The Skull and Crossbones Nebula and NGC 2453

The constellation Puppis the Stern also contains the Skull and Crossbones Nebula (NGC 2467), a nice emission region that lies 1.7° south-southeast of the 3rd-magnitude star Xi (ξ) Puppis. A 7th-magnitude open cluster with the same NGC number surrounds the nebula. Find them with binoculars before swinging a scope in their direction.

Deep images show the colorful emission nebula and drown out the 50-star cluster, but through a small telescope, you'll find a surprisingly bright, round nebulosity embedded in a rich star field. Recognizing anything akin to crossbones proves problematic, but the skull is there. Some observers report seeing the nebula in 10x50 binoculars. An OIII filter helps—the object lies far south, and the filter helps compensate for the atmospheric loss at this southernly declination.

NGC 2453 is a small, compressed open star cluster 1.3° southwest of NGC 2467. Because the Milky Way runs through most of Puppis, the constellation holds many more star clusters. You can spend several nights exploring them all, but we don't have the space to cover them in this book. And many lie too far south to see well from the northern United States.

The Skull and Crossbones Nebula (NGC 2467) is a small, round emission nebula in northern Puppis just 1.7° south-southeast of Xi Puppis. ESO/DSS2

The Skull and Crossbones Nebula
Also known as: NGC 2467
Constellation: Puppis the Stern
Right ascension: 7h52m
Declination: –26°27'
Magnitude: 7.1
Apparent size: 16'
Distance: 13,000 light-years

NGC 2453
Constellation: Puppis the Stern
Right ascension: 7h48m
Declination: –27°12'
Magnitude: 8.3
Apparent size: 5'
Distance: 8,000 light-years

SPRING SKY

■ Spring in the Northern Hemisphere brings warmer days and warmer nights. It also offers observers a new class of celestial objects to visit: galaxies beyond the Milky Way. Many observers refer to spring as "galaxy season" because our view at this time of year looks away from the dust that fills the Milky Way's disk and into relatively unobscured space. Coincidentally and fortuitously, the closest large galaxy cluster to Earth lies in this region. The Virgo Cluster of galaxies has the potential to keep you occupied for many nights.

This spring sky section covers objects in the night sky with right ascensions between approximately 8 and 14 hours. This quadrant of the sky places the constellation Leo the Lion above the horizon every night at 10 P.M. local daylight time from roughly March to June.

The spring sky doesn't contain the abundance of bright stars that winter holds. Still, you can find a few bright signposts. The most conspicuous of these is the Big Dipper, a prominent asterism in the northern constellation Ursa Major the Great Bear. If you follow the arc of the dipper's handle, you'll come to magnitude 0.0 Arcturus—the night sky's fourth-brightest star—in Boötes the Herdsman and then Spica, the luminary of Virgo the Maiden.

Although the spring sky may seem boring to casual naked-eye observers, at least compared with the wonders of winter, it is the prime season for breaking out a telescope and seeking distant galaxies. I've selected brighter island universes that show up through an 80mm refractor, but there's no denying you'll get a better view with a larger telescope. This is the one season where a big scope probably will be worth the higher cost and the extra hassle of hauling a big and heavy instrument to a dark-sky site.

My Top Ten list for spring observing, in order of total brightness, is: **1** Omega Centauri (NGC 5139), the sky's brightest globular cluster, in Centaurus; **2** globular cluster M3 in Canes Venatici; **3** Bode's Galaxy (M81) in Ursa Major; **4** spiral galaxy Centaurus A (NGC 5128) in Centaurus; **5** The Southern Pinwheel Galaxy (M83) in Hydra; **6** The Pinwheel Galaxy (M101) in Ursa Major; **7/8** [a tie] The Whirlpool Galaxy (M51) in Canes Venatici; **7/8** The Cigar Galaxy (M82) in Ursa Major; **9** the core of the Virgo galaxy cluster; and **10** The Leo Trio, a group of three adjacent galaxies in Leo.

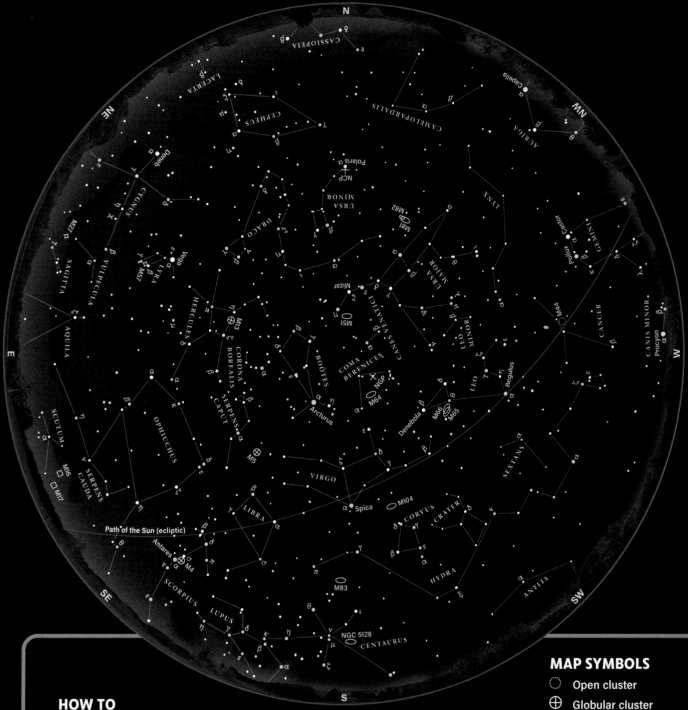

HOW TO USE THIS MAP

This map portrays the sky as seen near 35° north latitude. Located inside the border are the cardinal directions and their intermediate points. To find stars, hold the map overhead and orient it so one of the labels matches the direction you're facing. The stars above the map's horizon now match what's in the sky.

The all-sky map shows how the sky looks at:

1 A.M. March 15
11 P.M. April 15
9 P.M. May 15

STAR COLORS

A star's color depends on its surface temperature.

- The hottest stars shine blue
- Slightly cooler stars appear white
- Intermediate stars (like the Sun) glow yellow
- Lower-temperature stars appear orange
- The coolest stars glow red
- Fainter stars can't excite our eyes' color receptors, so they appear white unless you use optical aid to gather more light

MAP SYMBOLS

- ⬚ Open cluster
- ⊕ Globular cluster
- ▢ Diffuse nebula
- ✧ Planetary nebula
- ◯ Galaxy

STAR MAGNITUDES

- ● Sirius
- ● 0.0
- ● 1.0
- ● 2.0
- · 3.0
- · 4.0
- · 5.0

PAGE	OBJECT NAME	CONSTELLATION	TYPE	R.A.	DEC.	MAG.	SIZE
56	M48 (NGC 2548)	Hydra	Open cluster	8h14m	−5°45'	5.8	30'
57	NGC 2903	Leo	Spiral galaxy	9h32m	21°30'	8.9	13'x7'
58-59	The Beehive Cluster (M44)	Cancer	Open cluster	8h40m	19°40'	3.1	95'
58-59	M67 (NGC 2682)	Cancer	Open cluster	8h51m	11°49'	6.9	30'
60-61	Bode's Galaxy (M81)	Ursa Major	Spiral galaxy	9h56m	69°04'	6.9	21'x10'
60-61	The Cigar Galaxy (M82)	Ursa Major	Spiral galaxy	9h56m	69°41'	8.4	9'x4'
62	The Spindle Galaxy (NGC 3115)	Sextans	Lenticular galaxy	10h05m	−7°43'	9.9	8'x4'
63	The Ghost of Jupiter (NGC 3242)	Hydra	Planetary nebula	10h25m	−18°39'	7.8	25"
64-65	M95 (NGC 3351)	Leo	Spiral galaxy	10h44m	11°42'	9.7	7'x5'
64-65	M96 (NGC 3368)	Leo	Spiral galaxy	10h47m	11°49'	9.2	7'x5'
64-65	M105 (NGC 3379)	Leo	Elliptical galaxy	10h48m	12°35'	9.3	5.1'x4.7'
66-67	The Owl Nebula (M97)	Ursa Major	Planetary nebula	11h15m	55°01'	9.9	3.4'
66-67	M108 (NGC 3556)	Ursa Major	Spiral galaxy	11h12m	55°40'	10.1	8'x2'
68-69	M65 (NGC 3623)	Leo	Spiral galaxy	11h19m	13°06'	9.3	8'x2'
68-69	M66 (NGC 3627)	Leo	Spiral galaxy	11h20m	12°59'	8.9	9'x4'
68-69	The Hamburger Galaxy (NGC 3628)	Leo	Spiral galaxy	11h20m	13°35'	9.5	14'x4'
70	M106 (NGC 4258)	Canes Venatici	Spiral galaxy	12h19m	47°18'	8.3	19'x7'
71	M87 (NGC 4486)	Virgo	Elliptical galaxy	12h31m	12°23'	8.6	7'x7'
71	M84 (NGC 4374)	Virgo	Elliptical galaxy	12h25m	12°53'	9.2	7'x6'
71	M86 (NGC 4406)	Virgo	Elliptical galaxy	12h26m	12°57'	8.9	9'x6'
72	The Cocoon Galaxy (NGC 4490)	Canes Venatici	Spiral galaxy	12h31m	41°39'	9.4	7'x2'
72	NGC 4449	Canes Venatici	Irregular galaxy	12h28m	44°06'	9.4	6'x4'

PAGE	OBJECT NAME	CONSTELLATION	TYPE	R.A.	DEC.	MAG.	SIZE
73	The Sombrero Galaxy (M104)	Virgo	Spiral galaxy	12h40m	–11°37'	8.0	9'x4'
73	The Whale Galaxy (NGC 4631)	Canes Venatici	Spiral galaxy	12h42m	32°32'	9.8	17'x4'
74	The Blackeye Galaxy (M64)	Coma Berenices	Spiral galaxy	12h57m	21°41'	8.5	9'x5'
75	Centaurus A (NGC 5128)	Centaurus	Elliptical galaxy	13h25m	–43°01'	7.0	26'x18'
76	Omega Centauri (NGC 5139)	Centaurus	Globular cluster	13h27m	–47°29'	3.5	36'
77	M3 (NGC 5272)	Canes Venatici	Globular cluster	13h42m	28°23'	6.4	18'
78-79	The Whirlpool Galaxy (M51)	Canes Venatici	Spiral galaxy	13h30m	47°14'	8.4	11'x7'
80-81	The Southern Pinwheel (M83)	Hydra	Spiral Galaxy	13h37m	–29°52'	7.5	14'x13'
80-81	The Pinwheel Galaxy (M101)	Ursa Major	Spiral galaxy	14h03m	54°21'	7.9	29'x 27'

Open cluster M48

Near the far western boundary of the massive constellation Hydra the Water Snake lies the magnificent open star cluster M48 (NGC 2548). Hydra has many deep-sky treasures to explore, and we will feature a few of them, but most are distant galaxies better suited for large telescopes. Still, the brighter objects pack a solid punch for those with small scopes.

For a Messier object, M48 often gets overlooked. Maybe that's because the winter sky offers so many wonderful open clusters along the Milky Way, and observers simply are eager to tackle the spectacular galaxies in the spring sky. M48 shines at magnitude 5.8, so you should be able to pick it out with your naked eye under a dark sky. Binoculars show the cluster nicely.

The easiest way to find M48 is to start at magnitude 0.4 Procyon in Canis Minor and then scan 14° (about two binocular fields of view) southeast. Once you study M48 with binoculars, point your telescope at the bright object. This large, beautiful, and rich open cluster shows up easily in an 80mm refractor with a 22mm wide-field eyepiece. You'll clearly see its 80 or so stars spread across a diameter of 30' (the size of a Full Moon). A trio of 13th- and 14th-magnitude galaxies lurks 2.5° east-northeast of M48, but they lie well beyond the capabilities of an 80mm scope.

M48 lies some 2,500 light-years from Earth, which gives it a physical diameter of about 22 light-years. Astronomers estimate that it is roughly 400 to 500 million years old. That places it at an intermediate age between the younger Pleiades and older Hyades clusters in Taurus.

M48

Also known as: NGC 2548

Constellation: Hydra the Water Snake

Right ascension: 8h14m

Declination: –5°45'

Magnitude: 5.8

Apparent size: 30'

Distance: 2,500 light-years

The naked-eye open star cluster M48 lies in western Hydra near its border with Monoceros. A wonderful cluster that can be enjoyed with binoculars or small telescopes, M48 holds nearly 100 stars and resides about 2,500 light-years from Earth. NOIRLAB/NSF/AURA

Spiral galaxy NGC 2903

Directly north of one small segment of Hydra (it is, after all, the largest constellation and the one with the greatest east-west extent) lies galaxy-rich Leo the Lion. With binoculars (preferably tripod mounted), look 1.5° south of Lambda (λ) Leonis, a magnitude 4.3 star that stands just west of Leo's Sickle asterism. There you'll find a tiny non-stellar object on which to train your telescope: the barred spiral galaxy NGC 2903. This island universe resides about 30 million light-years from Earth.

You can see NGC 2903 with direct vision through an 80mm telescope, although it does not show much detail. You should be able to discern a brighter central bulge and fainter disk that appears elongated. NGC 2903 is famous among amateur astronomers as one of the few galaxies that visually shows spiral arm detail in larger instruments. You may also detect a stellar nucleus, but you likely will need at least an 8-inch scope for that.

In the 19th century, William Parsons, the Third Earl of Rosse, discerned this galaxy's spiral arms through his metal-mirrored 72-inch telescope, the Leviathan of Parsonstown. I often wonder if his youngest son, Sir Charles, gazed upon those spinning galactic arms thinking they were some sort of cosmic engine. Sir Charles inherited his father's engineering talent and later went on to invent the steam turbine.

Many observers wonder why Charles Messier did not find or record this galaxy for his catalog. At magnitude 8.9, it's at least as bright of many of the 9th- and 10th-magnitude objects the French astronomer recorded. I also wonder why Patrick Moore did not include NGC 2903 in his Caldwell Catalog. After all, several of those entries prove more difficult to see.

I suspect Moore simply ran out of room on his list of 109 targets beyond the Messier Catalog. I could have added many other worthy objects to this book, but space was limited, so I understand Moore's limitations.

NGC 2903
Constellation: Leo the Lion
Right ascension: 9h32m
Declination: 21°30′
Magnitude: 8.9
Apparent size: 13′x7′
Distance: 30 million light-years

NGC 2903 in Leo shines at 9th magnitude, making it the constellation's second-brightest galaxy—brighter than four of the Lion's Messier galaxies. R. JAY GABANY

The Beehive Cluster (M44) is a large naked-eye open star cluster in the constellation Cancer. Located some 515 light-years away, it is one of the closest star clusters to Earth.

NOIRLAB/NSF/AURA

Star clusters in Cancer

Look near the center of the constellation Cancer the Crab some 12° due west of NGC 2903 and you'll land on the huge Beehive star cluster (M44). Because this rich cluster lies only about 515 light-years from Earth, it appears both big (1.6° in diameter) and bright (magnitude 3.1). In fact, the lackluster constellation Cancer is the only one out of all 88 populating the sky that has a deep-sky object brighter than any of its stars.

Easy to see with the naked eye, the Beehive looks best through binoculars and small scopes operating at low power. This is one of the few objects in this book that you can enjoy in moderately bright urban environments or when the Full Moon is out (though the view does suffer from the excess light).

Big telescopes and high powers don't work well on the Beehive because they won't let you see the entire cluster at once. Large

scopes provide only one advantage: They might let you track down the six 13th- to 15th-magnitude galaxies hiding among the Beehive's stars. Five of these background island universes have their own NGC numbers.

At least 1,000 stars crowd together in M44's 15-light-year-diameter volume. Sharp-eyed observers have seen several of these suns without optical aid, but most people need a bit of help.

The Beehive Cluster
Also known as:
Praesepe, M44,
NGC 2632
Constellation: Cancer the Crab
Right ascension: 8h40m
Declination: 19°40'
Magnitude: 3.1
Apparent size: 95'
Distance: 515 light-years

Astronomers think the cluster formed about 600 million years ago. And researchers have found at least two exoplanets orbiting stars within the Beehive. That's something to consider as you gaze upon this spring gem.

Scan 8.3° to the south-southeast from the Beehive and you'll discover Cancer's second Messier object: M67 (NGC 2682) This open cluster lies 1.6° due west of magnitude 4.3 Alpha (α) Cancri. You won't see this object with your naked eye because it shines at magnitude 6.9, but it looks wonderful through binoculars and small telescopes. M67 spans 30', so it appears as big as the Full Moon. The cluster contains about 500 stars, one of which shines with a noticeable red color, so some observers have dubbed it the King Cobra or Golden Eye cluster. Can you see the golden eye?

Astronomers estimate M67 to be perhaps 5 billion years old, much older than the Beehive. Scientists use star clusters to test their theories of how stars evolve because all the cluster members formed at the same time and with roughly the same chemical composition.

Although fainter, older, and more distant than M44, M67 still shows up through binoculars. It appears one-third the size of its fellow Messier object in Cancer. NOIRLAB/NSF/AURA

M67
Also known as: NGC 2682
Constellation: Cancer the Crab
Right ascension: 8h51m
Declination: 11°49'
Magnitude: 6.9
Apparent size: 30'
Distance: 2,800 light-years

Great galaxies in the Great Bear

Spiral galaxies M81 (left) and M82 (right) in northern Ursa Major form a mesmerizing pair through binoculars and small telescopes. GIUSEPPE DONATIELLO/ CC0 1.0 UNIVERSAL (CC0 1.0)

Ursa Major the Great Bear swings high in the north on spring evenings. Like Leo, this constellation holds lots of galaxies for backyard observers. It's easy to compile a list of more than 30 NGC galaxies worth exploring in Ursa

Major; chasing them all down makes for a pleasant night or two of viewing. To find your way around the Great Bear, use its famous asterism—the Big Dipper— as a guide.

Our first two objects lie in far northern Ursa Major.

To find Bode's Galaxy (M81) and the Cigar Galaxy (M82) with binoculars, draw an imaginary line from magnitude 2.4 Gamma (γ) Ursae Majoris (the star at the southeastern corner of the dipper's bowl) through magnitude 1.8

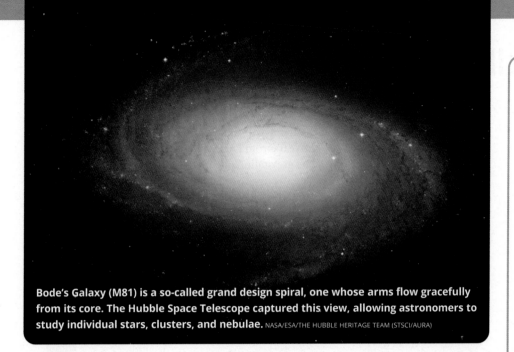

Bode's Galaxy (M81) is a so-called grand design spiral, one whose arms flow gracefully from its core. The Hubble Space Telescope captured this view, allowing astronomers to study individual stars, clusters, and nebulae. NASA/ESA/THE HUBBLE HERITAGE TEAM (STSCI/AURA)

Bode's Galaxy
Also known as: M81, NGC 3031
Constellation: Ursa Major the Great Bear
Right ascension: 9h56m
Declination: 69°04'
Magnitude: 6.9
Apparent size: 21'x10'
Distance: 12 million light-years

The Cigar Galaxy
Also known as: M82, NGC 3034
Constellation: Ursa Major the Great Bear
Right ascension: 9h56m
Declination: 69°41'
Magnitude: 8.4
Apparent size: 9'x4'
Distance: 12 million light-years

Alpha (α) Ursae Majoris (at the bowl's northwestern corner) and extend it by an equal length. Binoculars show the galaxies well, but they stand out through a telescope. They have to rank as one of the nicest galaxy pairs in the sky.

The galaxies glow at magnitudes 6.9 (M81) and 8.4 (M82). Although a few people with great eyesight have detected M81 with the unaided eye under excellent skies, you need optical aid to do them justice. Both galaxies are easy targets through an 80mm refractor. Bode's Galaxy appears oval with a fairly bright central bulge and an obviously less-populated disk (the spiral arms). The galaxy shows definite spiral structure in large amateur instruments. German astronomer Johann Bode discovered M81 in 1774; Messier added it to his catalog in 1779.

M82 appears far different, even in binoculars. It looks more elongated and has rounded ends, indeed much like a short cigar. Under extremely good conditions, or by using an

8-inch or larger telescope, you may see dust lanes cutting across M82's disk in one or more locations.

Look 0.8° east of M81 and you'll find the magnitude 9.9 galaxy NGC 3077. An 80mm wide-field telescope at low power reveals

this satellite galaxy as a featureless oval glow. Another satellite, magnitude 10.1 NGC 2976, lies 1.7° southwest of M81. You'll have a hard time detecting this galaxy with an 80mm scope unless you have excellent skies.

The Cigar Galaxy (M82) is a starburst galaxy creating stars at a rate 10 times that of the Milky Way. Notice the plumes of glowing hydrogen erupting from its central regions. NASA/ESA/THE HUBBLE HERITAGE TEAM (STSCI/AURA)

The Spindle Galaxy

Head south from Ursa Major through Leo Minor and Leo and you'll come to the faint constellation Sextans the Sextant. Locate magnitude 5.1 Gamma (γ) Sextantis and then slide 3.2° east to our next target: the Spindle Galaxy (NGC 3115). Although some observers see it as non-stellar with binoculars, you really need a telescope to confirm its identity.

NGC 3115 shows up easily with direct vision through an 80mm-diameter telescope. I recommend you try to acquire the galaxy with an 80mm or larger scope at low power and then pop in a higher-power eyepiece once you have it centered in the eyepiece. Galaxies typically look better at higher powers unless you're viewing an unusually large one or observing a group or cluster.

At first glance, the Spindle Galaxy could pass for an edge-on spiral similar to the Cigar Galaxy (M82). But NGC 3115 comes with a twist—or more properly, a lump. A pronounced bulge lies at its center, making the galaxy look like the cross section of a lens. Such galaxies are called "lenticular." Essentially, they have spiral shapes but lack the gas needed to make new stars. An 80mm refractor reveals this basic shape.

The Spindle Galaxy lurks about 32 million light-years from Earth and seems gravitationally unattached to any other galaxy. Astronomers have discovered a supermassive black hole weighing at least 1 billion solar masses at the heart of NGC 3115. Observations made with the European Southern Observatory's Very Large Telescope and the Chandra X-Ray Observatory have produced images showing superheated gas flowing into the galaxy's black hole, the first time this behavior had been observed.

The Spindle Galaxy
Also known as: NGC 3115, Caldwell 53
Constellation: Sextans the Sextant
Right ascension: 10h05m
Declination: –7°43'
Magnitude: 9.9
Apparent size: 8'x4'
Distance: 32 million light-years

The Spindle Galaxy (NGC 3115) shows a thin profile with a central bulge typical of edge-on systems. NASA/ESA/ J. ERWIN (UNIVERSITY OF ALABAMA)

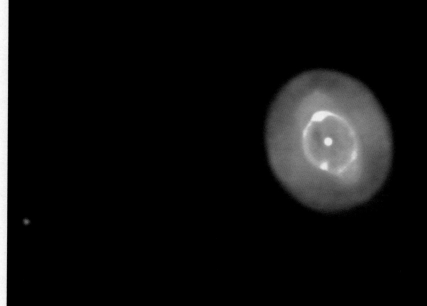

The Ghost of Jupiter

Hydra the Water Snake lies due south of Sextans, making our next target an easy hop from the Spindle Galaxy. The Ghost of Jupiter (NGC 3242) lies 11.9° south-southeast of the Spindle, though it's easier to find by sliding 1.8° south of magnitude 3.8 Mu (μ) Hydrae. The Ghost of Jupiter stands as the finest planetary nebula in the spring sky. Although some observers claim to have seen NGC 3242 with binoculars, the 25"-diameter object looks like a star. Under excellent conditions with large binoculars, the "star" might show a subtle blue or blue-green color.

At magnitude 7.8, the Ghost of Jupiter glows brightly enough to see through an 80mm telescope. Unfortunately, it's so small as to appear starlike at low powers. Crank up the power to confirm your sighting of a nebulous object. If Jupiter happens to be visible, you might want to take a look at it with the same telescope-eyepiece combination. The nebula is only slightly smaller than the planet.

Oddly enough, the Ghost of Jupiter more closely mimics the color and brightness of Uranus and Neptune. Once you center it in your eyepiece, the nebula appears as a small, nearly circular ball that glows blue-green. The color comes from atoms of doubly ionized oxygen (OIII) in the planetary's shell. These ions absorb ultraviolet light emanating from the hot central star and then reradiate it in the visible spectrum. As you might guess, you can pump up the contrast on NGC 3242 by using an OIII filter, though the object shines brightly enough not to require it. You'll need a 12-inch or larger telescope to see the central star. Even bigger scopes show the object as a bluish disk with a white central star.

The Ghost of Jupiter serves as a perfect specimen showing how planetary nebulae got their name—they strongly resemble our solar system's outer planets when viewed through a small telescope. But planetaries actually represent the end stages of a star much like the Sun. After exhausting its hydrogen and helium fuel, the star becomes unstable and puffs off its outer layers.

The Ghost of Jupiter (NGC 3242) in the constellation Hydra is a bright planetary nebula, not much smaller than the size of Jupiter when viewed through a telescope. ADAM BLOCK

The Ghost of Jupiter

Also known as: The Eye Nebula, NGC 3242, Caldwell 59

Constellation: Hydra the Water Snake

Right ascension: 10h25m

Declination: –18°39'

Magnitude: 7.8

Apparent size: 25"

Distance: 4,800 light-years

M95
Also known as: NGC 3351
Constellation: Leo the Lion
Right ascension: 10h44m
Declination: 11°42'
Magnitude: 9.7
Apparent size: 7'x5'
Distance: 33 million light-years

M96
Also known as: NGC 3368
Constellation: Leo the Lion
Right ascension: 10h47m
Declination: 11°49'
Magnitude: 9.2
Apparent size: 7'x5'
Distance: 31 million light-years

M105
Also known as: NGC 3379
Constellation: Leo the Lion
Right ascension: 10h48m
Declination: 12°35'
Magnitude: 9.3
Apparent size: 5.1'x4.7'
Distance: 37 million light-years

Neighbors of a different stripe

Head back to Ursa Major the Great Bear and its conspicuous asterism, the Big Dipper. Concentrate on the two 2nd-magnitude stars that form the bottom of the dipper's bowl: Beta (β) and Gamma (γ) Ursae Majoris. Our next targets lie about one-quarter of the way from Beta to Gamma and just south of the line connecting them. Can you spot the Owl Nebula (M97) and galaxy M108 (NGC 3556) with just binoculars?

Aim your wide-field telescope with a low-power eyepiece here and you will get a lovely vista including both objects. The two targets show up in an 80mm refractor, but they are not easy, and you should look for them when they lie near the meridian. Both rank among the faintest objects in the Messier Catalog.

M108 shines at magnitude 10.1. It appears as an almost linear glow because it is a nearly edge-on spiral galaxy. With superb seeing and transparency you might see some mottling of the galaxy's disk, but this observation comes close to the limit of what an 80mm telescope can achieve.

The Owl Nebula covers a lot of territory for a planetary. (Its diameter measures about eight times that of the Ghost of Jupiter.) The nebula's size means its light spreads out over a large area, giving it a low surface brightness and making it difficult to see. Like the Ghost of Jupiter, the Owl responds well to an OIII filter, which makes the nebula more noticeable. Two apparent holes—the Owl's "eyes"—mar M97's disk. You may see the Owl's face with the help of an OIII filter, but it usually requires a 10-inch telescope.

M97 resides 2,600 light-years from Earth while M108 lies 46 million light-years away. Their alignment occurs by chance.

Spiral galaxy M108 in Ursa Major appears nearly edge-on from our perspective. At 10th magnitude, it requires a clear dark sky to see through an 80mm telescope. NOIRLAB/NSF/AURA

The Leo Triplet

Journey down south to Leo the Lion once more, to the constellation's eastern end. Find the magnitude 3.3 star Theta (θ) Leonis and then scan 2.5° southeast with your binoculars. You can't miss the Leo Triplet—the galaxies M65 (NGC 3623), M66 (NGC 3627), and the Hamburger Galaxy (NGC 3628). All three show up nicely in binoculars and with direct vision through an 80mm refractor. M65 and M66 form arguably the second-best galaxy pair in the northern sky, after M81 and M82 in Ursa Major. They lie 20' apart. The Hamburger Galaxy stands 36' north of the other two.

All three are spiral galaxies. We see M65 tipped strongly to our line of sight. It shows a noticeable disk and a bright core. M66 appears more face-on and reveals a nice halo surrounding its bright core. Edge-on spiral NGC 3628 glows fainter than the other two. You may detect a hint of the dark lane running through the galaxy's disk, which gives the Hamburger its name, but many observers say you need a 16-inch scope to see it clearly.

The distorted spiral arms of M66 (NGC 3627) tell the tale of gravitational interactions with its neighbors, M65 and NGC 3628. KPNO/ NOIRLAB/NSF/AURA/JEFF HAPEMAN/ADAM BLOCK

A second trio of galaxies lies in eastern Leo, even brighter than the M96 group. Here you'll find M65 (top right), M66 (bottom right), and the Hamburger Galaxy (left): JOEL SHORT

M65 (NGC 3623) turns nearly edge-on to our line of sight, giving observers a nice opportunity to see its dusty disk. NOIRLAB/NSF/AURA

M65
Also known as: NGC 3623
Constellation: Leo the Lion
Right ascension: 11h19m
Declination: 13°06'
Magnitude: 9.3
Apparent size: 8'x2'
Distance: 35 million light-years

M66
Also known as: NGC 3627
Constellation: Leo the Lion
Right ascension: 11h20m
Declination: 12°59'
Magnitude: 8.9
Apparent size: 9'x4'
Distance: 36 million light-years

The Hamburger Galaxy
Also known as: NGC 3628
Constellation: Leo the Lion
Right ascension: 11h20m
Declination: 13°35'
Magnitude: 9.5
Apparent size: 14'x4'
Distance: 35 million light-years

The aptly named Hamburger Galaxy (NGC 3628) tilts edge-on to our line of sight, giving us a good view of its distorted dust lane. KPNO/NOIRLAB/NSF/AURA/ GLEN SAURDIFF AND JOAN SIMPSON/F. HAASE, S. PETERSON, K. GARMANY

Spiral galaxy M106

The barred spiral galaxy M106 (NGC 4258) lies in Canes Venatici just south of Ursa Major. Visible in binoculars, M106 hosts an active galactic nucleus powered by a supermassive black hole. KPNO/NOIRLAB/NSF/AURA

Swing south of the Big Dipper's handle and you'll land in the nondescript but galaxy-rich constellation Canes Venatici the Hunting Dogs. Our next target lies in this constellation's north-western corner. To find the barred spiral galaxy M106 (NGC 4258), start at the magnitude 2.4 star Gamma (γ) Ursae Majoris at the southeastern corner of the dipper's bowl. Then scan with binoculars 6° south and pick up magnitude 3.7 Chi (χ) Ursae Majoris. M106 lies 5.6° due east of Chi. Glowing at magnitude 8.3, this bright galaxy shows up nicely through binoculars and is easy to see with an 80mm telescope and 22mm eyepiece.

M106 lies about 24 million light-years from Earth, making it one of the closer galaxies outside our Local Group. It also spans 135,000 light-years, making it slightly larger than the Milky Way. Its size and distance conspire to give it a decent size of 19'x7'.

M106 isn't quite normal, however. Astronomers classify it as a Seyfert galaxy—which means it has an active core and an unusual spectrum with narrow emission lines. American astronomer Carl Seyfert first studied such galaxies in the early 1940s. The strong activity in M106's core comes from a super-massive black hole of some 30 million solar masses

that actively swallows gas and dust in its vicinity. The galaxy also hosts a so-called megamaser, meaning it emits coherent microwave light visible to radio telescopes instead of visible light like lasers.

Through an 80mm telescope, the bright nucleus dominates the view. You also can see the galaxy's long oval structure, which comes courtesy of its 64° tilt to our line of sight, and its uneven brightness. More than a dozen other galaxies lie within a 4° circle centered on M106, but the brightest glows dimly at 11th magnitude, making it a difficult object with an 80mm telescope unless you image it.

M106
Also known as: NGC 4258
Constellation: Canes Venatici the Hunting Dogs
Right ascension: 12h19m
Declination: 47°18'
Magnitude: 8.3
Apparent size: 19'x7'
Distance: 24 million light-years

The heart of the Virgo Cluster

M87

Also known as: NGC 4486

Constellation: Virgo the Maiden

Right ascension: 12h31m

Declination: 12°23'

Magnitude: 8.6

Apparent size: 7'x7'

Distance: 55 million light-years

M84

Also known as: NGC 4374

Constellation: Virgo the Maiden

Right ascension: 12h25m

Declination: 12°53'

Magnitude: 9.2

Apparent size: 7'x6'

Distance: 55 million light-years

M86

Also known as: NGC 4406

Constellation: Virgo the Maiden

Right ascension: 12h26m

Declination: 12°57'

Magnitude: 8.9

Apparent size: 9'x6'

Distance: 55 million light-years

The core of the Virgo Cluster of galaxies is a scene like no other. The biggest and brightest galaxy here is M87 (NGC 4486). The other two brighter galaxies are M84 (NGC 4374) and M86 (NGC 4406). All three can be seen in larger binoculars. RICH JACOBS

Southeast of Leo the Lion lies the second-largest constellation in the sky: Virgo the Maiden. And it features one of the most spectacular regions visible in spring or any other season—the core of the massive Virgo Cluster of galaxies. The cluster covers so much sky that it bleeds into the neighboring constellation Coma Berenices.

Start with magnitude 2.1 Denebola, the tail star of Leo, then pick up magnitude 2.8 Epsilon (ε) Virginis 18° to the east-southeast. Galaxies litter the area between these two stars. This is the heart of the Virgo Cluster, the closest large galaxy cluster to the Milky Way. More than 1,000 galaxies call this region home. And the cluster itself lies at the center of the vastly larger Virgo Supercluster, which includes our Local Group of galaxies and, thus, the Milky Way, near its outer edge.

With a wide-field telescope, point to the area halfway between Denebola and Epsilon Virginis. A quick sweep around with an 80mm telescope should reveal two or three fuzzy patches, and perhaps a half dozen or more show up through a 6-inch instrument. (The entire cluster holds roughly 150 galaxies visible in a 6-inch scope.)

The three brightest galaxies—M84 (NGC 4374), M86 (NGC 4406), and M87 (NGC 4486)—all glow at 9th magnitude and show up in a single field through an 80mm wide-field refractor. These three ellipticals all rank among the most massive galaxies in the local universe. M87 stands out as the biggest and brightest, and earned its place in astronomical history in 2019 when astronomers imaged the event horizon of its supermassive black hole.

If you start at M84, head east to M86, and then continue in a northeast direction, you'll be able to trace an arc of about a half-dozen galaxies called Markarian's Chain. Some of the chain's fainter galaxies lie near the magnitude limit of an 80mm refractor, so try to track them down when they stand close to the meridian.

Galaxies go to the dogs

If the Virgo Cluster has filled your heart and mind with elliptical galaxies, let's focus on some stranger galaxies. Return to Canes Venatici the Hunting Dogs south of the Big Dipper's handle. Find the constellation's two brightest stars—magnitude 2.9 Alpha (α) and magnitude 4.3 Beta (β) Canum Venaticorum—then point your telescope 0.7° west-northwest of Beta. You should see the Cocoon Galaxy (NGC 4490) at the center of your field. Although it glows rather faintly at magnitude 9.4, it shows up quite easily.

This teardrop-shaped barred spiral lies 25 million light-years from Earth. Tidal interactions with a small companion galaxy, NGC 4485, have heavily disrupted the galaxy. As

a result, the Cocoon has been whipped up into a high rate of star formation, what astronomers call a starburst galaxy.

Head 2.5° due north and slightly west of the Cocoon to look for a similarly bright galaxy, NGC 4449. Astronomers describe NGC 4449 as an irregular galaxy because it has little organizational structure. It appears roughly triangular

in shape, and you might be able to pick out its stellar nucleus. In many respects, it is a smaller version of the Large Magellanic Cloud, a satellite galaxy of the Milky Way visible from the Southern Hemisphere. NGC 4449 resides about 13 million light-years away.

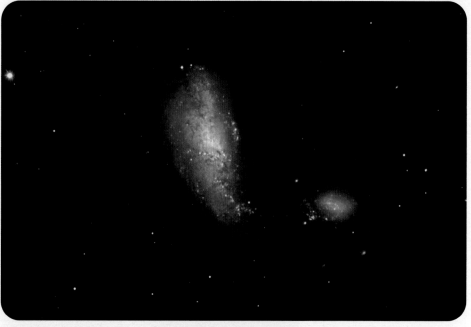

The Cocoon Galaxy (NGC 4490) actually consists of two interacting spiral galaxies. A tail of stars stretches between the two, which are now at least 24,000 light-years from each other. KPNO/NOIRLAB/NSF/AURA/MICHAEL GARIEPY/ADAM BLOCK

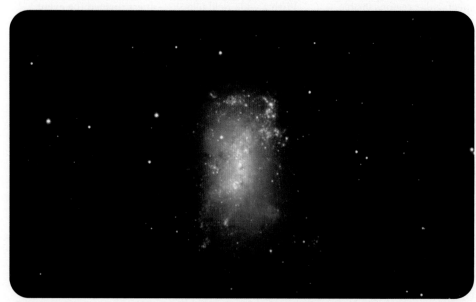

Irregular galaxy NGC 4449 in Canes Venatici resembles the Large Magellanic Cloud located deep in the southern sky. NGC 4449 contains lots of hydrogen gas that fuels a high rate of star formation. KPNO/NOIRLAB/NSF/AURA/JOHN AND CHRISTIE CONNORS/ADAM BLOCK

The Cocoon Galaxy
Also known as: NGC 4490
Constellation: Canes Venatici the Hunting Dogs
Right ascension: 12h31m
Declination: 41°39'
Magnitude: 9.4
Apparent size: 7'x2'
Distance: 25 million light-years

NGC 4449
Also known as: Caldwell 21
Constellation: Canes Venatici the Hunting Dogs
Right ascension: 12h28m
Declination: 44°06'
Magnitude: 9.4
Apparent size: 6'x4'
Distance: 13 million light-years

A Sombrero and a Whale

Another odd-looking but stunning galaxy lies in far southern Virgo the Maiden, just across the border from Corvus the Crow. You can find it most easily by scanning 11° due west of 1st-magnitude Spica. The Sombrero Galaxy (M104) lies nearly 30 million light-years from Earth but still manages to shine at magnitude 8.0. It's easy to pick up in binoculars and a cinch with a small telescope.

The Sombrero's name comes from its resemblance to the wide-brimmed Mexican hat. In M104's case, the shape derives from a prominent dust lane that crosses just south of the galaxy's center. The dust lane's angle arises because the galaxy tilts 6° to our line of sight. You might get a hint of the dust lane through an 80mm telescope at high power, but it usually takes a 6-inch or bigger instrument to see it clearly.

Next, work back to Canes Venatici the Hunting Dogs. Locate magnitude 2.9 Alpha (α) Canum Venaticorum and then scan 6.4° south-southwest. You'll be looking at the Whale Galaxy (NGC 4631), a magnitude 9.8 spiral residing 30 million light-years away. The Whale shows up through binoculars and 80mm refractors.

The Whale appears as a long, thin streak about half the Full Moon's size. Our edge-on view shows an asymmetry—it looks thicker at one end than the other, resembling a whale's silhouette.

Half a degree southeast of the Whale lies another edge-on galaxy: the Hockey Stick Galaxy (NGC 4656). This 10th-magnitude island universe seems to have a bent tip, which is really a small companion overlaid on the main galaxy's end. This galaxy really shines through large instruments.

The Sombrero Galaxy

Also known as: M104, NGC 4594
Constellation: Virgo the Maiden
Right ascension: 12h40m
Declination: –11°37'
Magnitude: 8.0
Apparent size: 9'x4'
Distance: 29 million light-years

The Whale Galaxy

Also known as: NGC 4631, Caldwell 32
Constellation: Canes Venatici the Hunting Dogs
Right ascension: 12h42m
Declination: 32°32'
Magnitude: 9.8
Apparent size: 17'x4'
Distance: 30 million light-years

At the southern edge of Virgo lurks the Sombrero Galaxy (M104). Images make it clear why: A dark band of dust slashing across the center looks like the brim of a hat. ESO/P. BARTHEL

The Whale Galaxy (NGC 4631) in Canes Venatici is a fine edge-on galaxy. You might see its small companion, NGC 4627, just north of it under excellent viewing conditions. KPNO/ NOIRLAB/NSF/AURA/JOHN VICKERY AND JIM MATTHES/ADAM BLOCK

The Blackeye Galaxy

The Blackeye Galaxy (M64) in Coma Berenices has an unusually large dust cloud superimposed in front of its spiral arms. NOIRLAB/NSF/AURA

Did I mention spring is galaxy season? Our next target, the Blackeye Galaxy (M64), lies 11.3° south-southeast of the Whale in Coma Berenices, Berenice's Hair. Use 5th-magnitude 35 Comae Berenices as your guide. The galaxy stands 0.9° east-northeast of this star. You can find M64 with binoculars and it shows up easily with direct vision through an 80mm telescope.

From the image above, you can see that M64 has an odd, heavy dust lane on one side of its nucleus. British astronomer William Herschel discovered this dark marking in 1785 and compared it to a black eye, and the name stuck. You'll have a difficult time seeing the dust lane in an 80mm telescope, but you may get a hint of it at higher powers with averted vision.

No one has pinned down the Blackeye Galaxy's distance, with estimates ranging from 14 million to 24 million light-years. The spiral has an active nucleus typical of a Seyfert galaxy. Oddly enough, the stars and gas in M64's core and inner disk rotate in the opposite direction from material in its outer regions. Astronomers surmise that a billion or so years ago, the Blackeye collided with and ultimately assimilated a smaller companion. The two galaxies have long since merged, but the counter-rotating parts preserve the memories of the two progenitors.

The Blackeye Galaxy

Also known as: M64, NGC 4826

Constellation: Coma Berenices, Bernice's Hair

Right ascension: 12h57m

Declination: 21°41'

Magnitude: 8.5

Apparent size: 9'x5'

Distance: 24 million light-years

Elliptical galaxy Centaurus A

Centaurus A lies in the southern constellation Centaurus the Centaur. The galaxy's strange appearance and active nature trace back to a collision between a large elliptical galaxy and a smaller spiral. ESO/ESA/HUBBLE/NASA/DSS

Most of the deep-sky objects I cover in this book climb reasonably high in our Northern Hemisphere sky. But a couple of southern objects look so spectacular that I just couldn't pass them up. The first is a strange-looking galaxy best known as Centaurus A (NGC 5128). It got this moniker because it was the first and strongest radio source in the constellation Centaurus the Centaur. At a declination of –43°, NGC 5128 can't be seen from north of latitude 47° north. That leaves out Canadian observers and makes it close to impossible for observers in the northern United States to view it. But if you live in one of the southern states, Centaurus A certainly merits a look.

To find NGC 5128, start at 1st-magnitude Spica in Virgo and then pick up the two 3rd-magnitude stars Gamma (γ) Hydrae and Iota (ι) Centauri that form a chain running to the south. Centaurus A lies 6.4° south-southeast of Iota. The magnitude 7.0 galaxy shows up in binoculars under a dark sky and ranks as an easy target through an 80mm telescope.

At first you may think you're looking at an elliptical galaxy, but the situation is more complicated. A bit of magnification gives a hint of a broad dust lane that cuts the galaxy nearly in half. Deep images reveal jets of material shooting from both the galaxy's north and south poles at half the speed of light.

Centaurus A qualifies as an active galaxy, and the 55-million-solar-mass supermassive black hole at its core puts on quite a light show. The galaxy emits copious amounts of radiation at wavelengths ranging from the radio to X-rays. NGC 5128 also creates new stars at a phenomenal rate. Astronomers suspect the galaxy's odd look and behavior stem from the time when the main elliptical galaxy collided with and absorbed a smaller spiral galaxy perhaps 200 million years ago.

Centaurus A
Also known as: NGC 5128, Caldwell 77
Constellation: Centaurus the Centaur
Right ascension: 13h25m
Declination: –43°01'
Magnitude: 7.0
Apparent size: 26'x18'
Distance: 12 million light-years

Globular cluster Omega Centauri

If you thought Centaurus A would mark the southern extent of our deep-sky wanderings, wait until you try for Omega Centauri (NGC 5139)— Omega lies 4.5° south of NGC 5128. Why did I dive so far south for an object that climbs high enough to observe only from the southern tier of states? Because Omega Centauri is without a doubt the finest globular cluster in the sky.

At magnitude 3.5, NGC 5139 shines brightly enough to see with the naked eye, and it's an easy binocular and small scope target. Your biggest

obstacle to viewing it likely will be its location low in the sky, which will have you dodging trees or buildings to get a good look.

Omega Centauri got its name because early observers mistook it for a bright star. German cartographer Johann Bayer designated it Omega (ω) in his 1603 star atlas *Uranometria*. Edmond Halley gets credit for first noting its non-stellar appearance 74 years later.

This spectacular object is the brightest globular cluster in the sky. Part of that comes from its proximity—it lies only 17,000 light-years away—but more

of it stems from its abundance of stars. Astronomers estimate it holds up to 10 million stars packed in a ball some 150 light-years in diameter. That gives it a size of 36' on the sky—20 percent larger than the Full Moon. Astronomers suspect it to be the core of a dwarf galaxy the Milky Way assimilated long ago. On nights with steady air, an 80mm telescope will resolve stars on its edge while larger scopes resolve it across its face.

Another must-see object in Centaurus is the colossal globular cluster Omega Centauri. If you live where this object climbs above your southern horizon, it will become a favorite. ESO/INAF-VST/OMEGACAM

Omega Centauri
Also known as: NGC 5139, Caldwell 80
Constellation: Centaurus the Centaur
Right ascension: 13h27m
Declination: –47°29'
Magnitude: 3.5
Apparent size: 36'
Distance: 17,000 light-years

Globular cluster M3

The spring sky doesn't showcase many globular star clusters. Although Omega Centauri more than makes up for this deficit if you live far enough south, observers at more northerly locales will have to be satisfied with M3 (NGC 5272) in Canes Venatici the Hunting Dogs.

M3 sits at Canes Venatici's southern edge where it meets the constellation Boötes. To find the cluster with binoculars, start at magnitude 0.0 Arcturus, Boötes' brightest

star and the fourth brightest in the entire night sky. Then travel 12° northwest to M3. The cluster appears as a fairly bright fuzzball through binoculars and is a cinch to see with an 80mm telescope. At magnitude 6.4, M3 glows brightly enough that sharp-eyed observers can spot it with their naked eye from dark sites.

At low power in an 80mm scope, M3 appears as an unresolved circular object. Pump up the power and you will start to resolve stars at the cluster's edge.

The 14th-magnitude galaxy NGC 5263 lies 0.5° due west of M3, but you won't have a prayer of seeing it with 80mm of aperture.

Charles Messier stumbled upon M3 in 1764, the first Messier object he discovered himself. It lies some 33,000 light-years from Earth and holds about 500,000 stars in a ball measuring 180 light-years across. M3 also holds the record for most variable stars in a globular cluster, with 274 known at the time of this writing.

M3
Also known as: NGC 5272
Constellation: Canes Venatici the Hunting Dogs
Right ascension: 13h42m
Declination: 28°23'
Magnitude: 6.4
Apparent size: 18'
Distance: 33,000 light-years

The magnificent globular star cluster M3 lurks in southern Canes Venatici. It lies about twice as far from Earth as Omega Centauri and doesn't appear as spectacular, but it rises high in the northern spring sky and makes a rewarding target. T.A. RECTOR (UNIVERSITY OF ALASKA ANCHORAGE) AND H. SCHWEIKER (WIYN AND NOIRLAB/NSF/AURA)

Another celestial showpiece is the Whirlpool Galaxy (M51) in Canes Venatici. It is a classic grand design spiral galaxy with distinct spiral arms, the first galaxy to have spiral structure detected visually. NASA/ESA/S. BECKWITH (STSCI/THE HUBBLE HERITAGE TEAM (STSCI/AURA)

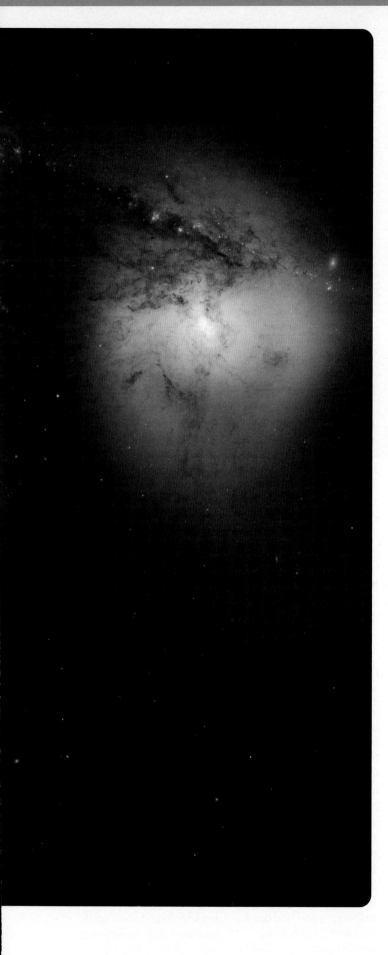

The Whirlpool Galaxy

One of the most iconic galaxies in the sky, the Whirlpool Galaxy (M51), lies just off the end star in the Big Dipper's handle. Begin your search with this star—magnitude 1.9 Eta (η) Ursae Majoris—and sweep 3.6° southwest with your binoculars. M51 shows up well through binoculars and small scopes.

Charles Messier discovered the Whirlpool in October 1773. He didn't detect any spiral structure, however. That fell to Lord Rosse, who first recorded the spiral arms on a drawing he made from observations with his 72-inch Leviathan Reflector in Ireland in 1845.

Long-exposure images with relatively modest telescopes today reveal an intricate, bold spiral pattern of striking symmetry and beauty that astronomers call a grand design spiral. Visually through an 80mm telescope, not so much. Oh, it's interesting as heck to find this object in the night sky and see the galaxy's major parts, but it isn't a blazing firework display that the images unveil. But it also doesn't cost thousands of dollars or require days of data gathering and image processing either. You can see it with your own eyes, and chase better views through bigger telescopes, an affliction amateur astronomers refer to as aperture fever.

Through an 80mm telescope, you will see a soft, even glow—the disk—surrounding a brighter core. Within this core is the active nucleus of a Seyfert galaxy. At one end of the round disk, M51 seems to blend into another round object. This is the satellite galaxy NGC 5195 that apparently passed through the outer disk of M51 nearly 100 million years ago and now lies on its far side. It takes an 8-inch telescope to see the spiral arms and a 12-inch or larger instrument to see them clearly.

The Whirlpool measures nearly 90,000 light-years across. (NGC 5195 spans some 55,000 light-years.) This makes M51 just slightly smaller than the Milky Way and the Andromeda Galaxy (M31; see the fall section). You won't want to miss out on this galaxy—it's one of the most intriguing objects visible from Earth.

The Whirlpool Galaxy

Also known as: M51, NGC 5194

Constellation: Canes Venatici the Hunting Dogs

Right ascension: 13h30m

Declination: 47°14'

Magnitude: 8.4

Apparent size: 11'x7'

Distance: 27 million light-years

Two giant pinwheels

The grand design spiral nicknamed the Southern Pinwheel (M83) in Hydra ranks among the sky's treasures. M83 lies 15 million light-years from Earth, making it one of the closest spirals to us. CTIO/NOIRLAB/DOE/NSF/AURA

Grand design spirals are nice to image but easier to view. Some 30° south of the celestial equator lies a remarkable example that many observers consider to be the southern twin to the Pinwheel Galaxy (M101). This magnitude 7.5 face-on spiral resides in Hydra the Water Snake, just north of that constellation's border with Centaurus.

It's an easy task to find M83 with binoculars. First locate magnitude 3.3 Pi (π) Hydrae, the brightest star at the eastern end of this mammoth constellation. Then pinpoint magnitude 2.1 Theta (θ) Centauri. M83 lies near the western apex of an imaginary equilateral triangle with Pi and Theta forming the base. You can use M83 and Hydra's second-best deep-sky showpiece, the Ghost of Jupiter (page 63), to get a sense of just how big this

constellation is: The two objects lie 45° apart and yet span less than half the Water Snake's total length.

Point your binoculars toward M83's position and you will see a soft glow. An 80mm telescope reveals a circular glow, the disk, along with a brighter core and the hint of a bar protruding from it. Alas, you won't see the spiral arms through an 80mm instrument, but a 6-inch telescope should reveal them.

M83 resides 15 million light-years from Earth and belongs to a modest group of galaxies that includes Centaurus A (page 75). The peculiar galaxy NGC 5253 lies 1.9° south-southeast of M83, though at magnitude 10.4 it likely lies beyond the reach of an 80mm scope. Astronomers think these two interacted in the past billion years, stoking star formation and making

both starburst galaxies. Observers have spotted six supernovae in M83 since 1923, but none since 1983. Are we overdue for another?

Our last spring object is a classic: the grand design spiral Pinwheel Galaxy (M101) in Ursa Major the Great Bear. To find M101 in binoculars, locate the last two stars in the Big Dipper's handle. Magnitude 1.9 Eta (η) Ursae Majoris stands at the handle's end while nearby magnitude 2.2 Zeta (ζ) Ursae Majoris marks the bend in the handle. (Zeta is also a well-known double star visible through binoculars.) Draw at imaginary line between Eta and Zeta and, from the midpoint, move northeast almost 5° (or about one field of view). Alternatively,

consider M101 the eastern apex of a nearly equilateral triangle with Eta and Zeta.

The Pinwheel shows up with 8x42 binoculars and proves to be an easy target through an 80mm telescope. Despite glowing at magnitude 7.9, its light spreads out over an area of 29'x27', or nearly the size of the Full Moon. This gives M101 a low surface brightness—the lowest of any Messier object. You'll need a dark sky and a clear night to see M101 well.

With an 80mm telescope, you'll see a relatively large, mostly featureless disk that appears quite symmetrical around a brighter core. Its stellar nucleus should show up if you use a little more power and wait until M101 lies

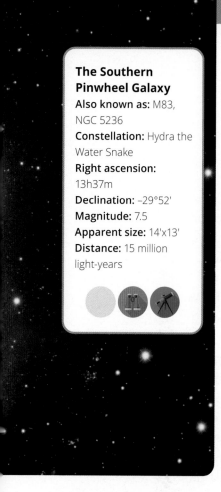
separate NGC numbers. Of these, NGC 5461, NGC 5462, and NGC 5471 are the most impressive. Astronomers using observatory telescopes have cataloged more than 1,200 emission nebulae in M101.

The Pinwheel Galaxy spans nearly 200,000 light-years, making it significantly larger than the Milky Way. It hosts roughly 1 trillion stars (a factor of three more than our home galaxy) and about 150 globular star clusters (close to the Milky Way's count).

Images reveal the galaxy as a classic grand design spiral oriented nearly face-on and with at least three distinct spiral arms. Many pinkish-red emission nebulae and giant bluish star clusters dot the disk, and a bright core seems to sit slightly off-center. This last feature arises because the galaxy has been interacting with some of its neighbors. The Pinwheel has lots of company: Nine other galaxies with NGC designations and 11 more too faint to be included in that catalog lie within 2° of M101. The brightest companion, NGC 5474, glows at 11th magnitude and stands 0.7° south-southeast of M101's center. This diminutive galaxy likely lies beyond the reach of an 80mm scope, but a 6-inch or larger instrument should have no trouble.

Can you spot any of the emission regions inside the Pinwheel or the companion galaxies surrounding it?

The Pinwheel Galaxy (M101) in Ursa Major makes a stunning spring showpiece. This spiral shows up through binoculars, while small scopes reveal its central bulge. You need a larger instrument to glimpse its spiral arms. T.A. RECTOR (UNIVERSITY OF ALASKA ANCHORAGE) AND H. SCHWEIKER (WIYN AND NOIRLAB/NSF/AURA)

near the meridian. With larger telescopes, you might glimpse some brighter areas in the galaxy's disk. These are emission nebulae, regions of glowing hydrogen gas, in a galaxy located 27 million light-years from Earth. Nine of them glow brightly enough that they were assigned

SUMMER SKY

▪ Summer means warm nights, vacations, and the opportunity to see the glories of the summer Milky Way if you travel to dark skies on one of your road trips. Plan your visit to a dark-sky site to coincide with the dark of the Moon (near New Moon) and bring a small telescope and binoculars to seek out some of nature's most spectacular night-sky vistas, or just to sweep along the Milky Way.

The summer sky section offers a selection of objects in the night sky with right ascensions between approximately 15 and 20 hours. This portion of the sky features the brightest portions of the Milky Way and the constellation Cygnus the Swan above the horizon every night at 10 P.M. local daylight time from roughly June to August.

Although the summer sky doesn't hold lots of 1st-magnitude stars, a conspicuous asterism will help you navigate around many of the Milky Way's treasures. The aptly named Summer Triangle comprises magnitude 0.0 Vega in Lyra the Harp—the night sky's fifth-brightest star—magnitude 0.8 Altair in Aquila the Eagle, and magnitude 1.3 Deneb in Cygnus the Swan.

The summer Milky Way looks so spectacular because we are peering toward the center of our galaxy. You can't actually see the center of the Milky Way Galaxy because the sheer number of stars and clouds of gas and dust obscures our view. I will point out several of the dark dust clouds and the bright emission nebulae that lie between Earth and our galaxy's core.

While summer nights are warmer, the evenings can get surprisingly cold—especially if you visit a national park located at high elevation. I have been caught in snowstorms in the high Sierra Mountains of California during August! Also remember that summer often means lots of stinging and biting bugs, so bring along insect repellent and lightweight, long-sleeve clothing.

My summer Top Ten list of can't-miss deep-sky objects, in order of overall brightness, is: **1** The False Comet in Scorpius; **2** Ptolemy's Cluster (M7), a sparkling open star cluster also in Scorpius; **3** the naked-eye globular star cluster M22 in Sagittarius; **4** the grand globular cluster M13 in Hercules; **5** the magnificent stretch of Milky Way from the Lagoon Nebula (M8) to the open star cluster M21 in Sagittarius; **6** the vista between the Eagle Nebula (M16) in Serpens and the Omega Nebula (M17) in Sagittarius; **7** the North America Nebula (NGC 7000) in Cygnus; **8** the Dumbbell Nebula (M27) in Vulpecula; **9** the Ring Nebula (M57) in Lyra; and **10** the naked-eye view of the summer Milky Way (granted, it's not the dimmest item on this list), which ranks among the most amazing sights in nature.

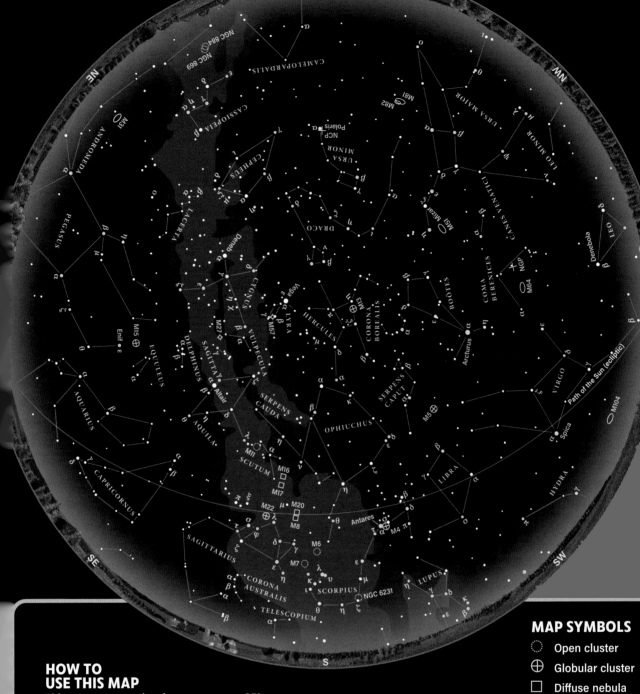

MAP SYMBOLS

◌ Open cluster

⊕ Globular cluster

▢ Diffuse nebula

✧ Planetary nebula

⬯ Galaxy

STAR MAGNITUDES

● Sirius

HOW TO USE THIS MAP

This map portrays the sky as seen near 35°
north latitude. Located inside the border are
the cardinal directions and their
intermediate points. To find stars, hold the
map overhead and orient it so one of the
labels matches the direction you're facing.
The stars above the map's horizon now
match what's in the sky.

The all-sky map shows how the sky looks at:

STAR COLORS

A star's color depends
on its surface temperature.

■ The hottest stars shine blue
■ Slightly cooler stars appear white
■ Intermediate stars (like the Sun) glow yellow

SUMMER SKY OBJECTS

PAGE	OBJECT NAME	CONSTELLATION	TYPE	R.A.	DEC.	MAG.	SIZE
86	M5 (NGC 5904)	Serpens	Globular cluster	15h19m	2°05'	5.8	23'
87	IC 4665	Ophiuchus	Open cluster	17h46m	5°39'	4.2	45'
88-89	M12 (NGC 6218)	Ophiuchus	Globular cluster	16h47m	−1°57'	6.7	16'
88-89	M10 (NGC 6254)	Ophiuchus	Globular cluster	16h57m	−4°06'	6.6	20'
88-89	M9 (NGC 6333)	Ophiuchus	Globular cluster	17h19m	−18°31'	7.7	9.3'
90-91	The Hercules Cluster (M13)	Hercules	Globular cluster	16h42m	36°28'	5.8	20'
90-91	NGC 6229	Hercules	Globular cluster	16h47m	47°32'	9.4	4.5'
90-91	M92 (NGC 6341)	Hercules	Globular cluster	17h17m	43°08'	6.4	14'
92-93	The Northern Jewel Box (NGC 6231)	Scorpius	Open cluster	16h54m	−41°50'	2.6	14'
92-93	Collinder 316 (Trumpler 24)	Scorpius	Open cluster	16h55m	−40°50'	6.4	60'
92-93	The Prawn Nebula (IC 4628)	Scorpius	Emission nebula	16h57m	−40°31'	—	90'x60'
94-95	The Pipe Nebula (Barnard 59, 65–7, 78)	Ophiuchus	Dark nebula	17h30m	−25°00'	—	500'x140'
94-95	The Snake Nebula (Barnard 72)	Ophiuchus	Dark nebula	17h24m	−23°42'	—	6'
96	The Butterfly Cluster (M6)	Scorpius	Open cluster	17h40m	−32°15'	4.2	25'
97	Ptolemy's Cluster (M7)	Scorpius	Open cluster	17h54m	−34°49'	3.3	80'
98-99	M21 (NGC 6531)	Sagittarius	Open cluster	18h04m	−22°30'	5.9	14'
98-99	The Trifid Nebula (M20)	Sagittarius	Emission nebula	18h03m	−22°58'	6.3	28'
98-99	NGC 6526	Sagittarius	Emission nebula	18h04m	−24°27'	8.0	40'
98-99	The Lagoon Nebula (M8)	Sagittarius	Emission nebula	18h04m	−24°23'	6.0	90'x40'
98-99	NGC 6544	Sagittarius	Globular cluster	18h07m	−25°00'	7.8	2'
98-99	NGC 6553	Sagittarius	Globular cluster	18h09m	−25°55'	8.1	8.2'
100	NGC 6520	Sagittarius	Open cluster	18h03m	−27°53'	7.6	5'
100	The Ink Spot (Barnard 86)	Sagittarius	Dark nebula	18h03m	−27°52'	—	5'
101	NGC 6541	Corona Australis	Globular cluster	18h08m	−43°43'	6.3	13'
101	The Emerald Nebula (NGC 6572)	Ophiuchus	Planetary nebula	18h12m	6°51'	8.1	14"
102	The Small Sagittarius Star Cloud (M24)	Sagittarius	Star cloud	18h17m	−18°31'	4.5	2°x1°
103	M18 (NGC 6613)	Sagittarius	Open cluster	18h20m	−17°06'	7.5	9'

PAGE	OBJECT NAME	CONSTELLATION	TYPE	R.A.	DEC.	MAG.	SIZE
104-105	NGC 6604/Sharpless 2-54	Serpens	Open cluster/ Emission nebula	18h18m	−12°15'	6.5	2'
104-105	The Eagle Nebula (M16)	Serpens	Emission nebula	18h19m	−13°48'	6.0	35'x28'
104-105	The Omega Nebula (M17)	Sagittarius	Emission nebula	18h21m	−16°10'	7.0	11'
106	The Tweedledum Cluster (NGC 6633)	Ophiuchus	Open cluster	18h27m	6°34'	4.6	27'
106	The Tweedledee Cluster (IC 4756)	Serpens	Open cluster	18h39m	5°28'	4.6	39'
107	The Wild Duck Cluster (M11)	Scutum	Open cluster	18h51m	−6°16'	5.3	14'
107	The Scutum Star Cloud	Scutum	Star cloud	18h40m	−10°00'	—	—
108	The Ring Nebula (M57)	Lyra	Planetary nebula	18h54m	33°02'	8.8	1.4'x1'
109	NGC 6755	Aquila	Open cluster	19h08m	4°16'	7.5	15'
109	NGC 6756	Aquila	Open cluster	19h09m	4°42'	10.6	4'
110	Brocchi's Cluster	Vulpecula	Asterism	19h25m	20°11'	3.6	60'
111	The Dumbbell Nebula (M27)	Vulpecula	Planetary nebula	20h00m	22°43'	7.4	8'x6'
112	Barnard's E (Barnard 142–3)	Aquila	Dark nebula	19h40m	10°31'	—	60'x40'
113	Barnard's Galaxy (NGC 6822)	Sagittarius	Irregular galaxy	19h45m	−14°48'	9.3	16'x14'
113	The Little Gem Nebula (NGC 6818)	Sagittarius	Planetary nebula	19h44m	−14°09'	9.4	22"x15"
114	NGC 6939	Cepheus	Open cluster	20h32m	60°39'	7.8	7'
114	The Fireworks Galaxy (NGC 6946)	Cygnus	Spiral galaxy	20h35m	60°09'	8.9	11'
115	The Veil Nebula (NGC 6960, NGC 6992/5)	Cygnus	Supernova remnant	20h56m	31°43'	7.0	210'x160'
116-117	The North America Nebula (NGC 7000)	Cygnus	Emission nebula	20h59m	44°20'	6.0	120'x100'
116-117	The Pelican Nebula (IC 5067/70)	Cygnus	Emission nebula	20h48m	44°22'	8.0	60'x50'
118-119	Gamma Cygni region	Cygnus	Emission nebula	20h19m	40°44'	10.0	50'
118-119	Albireo	Cygnus	Double star	19h31m	27°58'	3.1&5.1	34" separation
120-121	M22 (NGC 6656)	Sagittarius	Globular cluster	18h36m	−23°54'	5.1	32'
120-121	M28 (NGC 6626)	Sagittarius	Globular cluster	18h25m	−24°52'	6.8	11'
120-121	M69 (NGC 6637)	Sagittarius	Globular cluster	18h31m	−32°21'	7.6	11'
120-121	M70 (NGC 6681)	Sagittarius	Globular cluster	18h43m	−32°18'	7.9	8.0'
120-121	M54 (NGC 6715)	Sagittarius	Globular cluster	18h55m	−30°29'	7.6	12'
120-121	M55 (NGC 6809)	Sagittarius	Globular cluster	19h40m	−30°58'	6.3	19'
120-121	M75 (NGC 6864)	Sagittarius	Globular cluster	20h06m	−21°55'	8.5	6.8'

Globular cluster M5

Our first summer object, globular cluster M5 in Serpens the Serpent, ranks among the brightest globular star clusters in the sky. To find this cluster in binoculars, start at magnitude 2.6 Beta (β) Librae and head due north for 11° (about one-and-a-half binocular fields). M5 glows at magnitude 5.6, so it shows up to the naked eye if you observe under a dark sky. Needless to say, it's an easy object through binoculars and telescopes of any size.

As you focus on M5, you'll also see the 5th-magnitude star 5 Serpentis just 0.4° southeast of the cluster's center. Only five globulars outshine M5, and all of them lie south of the celestial equator. So it's surprising to find M5 so underappreciated. Perhaps the abundance of good globulars visible in late spring and early summer

has helped push this cluster to the back burner. M5 holds hundreds of thousands of stars and shines to us across 24,000 light-years.

In an 80mm telescope, you will see a distinctly fuzzy ball of stars in contrast to the background. At higher powers, you should start to resolve some stars around the cluster's edge.

Globular cluster M5 makes a fine binocular target, but it really stands out when viewed through a telescope of any size. STARHOPPER/CC BY-SA 4.0

M5
Also known as: NGC 5904
Constellation: Serpens the Serpent
Right ascension: 15h19m
Declination: 2°05'
Magnitude: 5.6
Apparent size: 23'
Distance: 24,000 light-years

Open cluster IC 4665

Open cluster IC 4665—seen here above Beta Ophiuchi—spans 0.75° and shows up best through binoculars and wide-field scopes. ALAN DYER

IC 4665 doesn't get much respect: It is the brightest star cluster not to make it into either the Messier or Caldwell catalog. In fact, IC 4665 shines brightly enough to be visible with the naked eye under a dark sky. The cluster lies in the northern part of Ophiuchus the Serpent-bearer.

To find IC 4665, first locate the constellation's fourth-brightest star, magnitude 2.8 Beta (β) Ophiuchi. Then point your binoculars or a wide-field telescope 1.3° to the northeast. There you will see a clump of fairly bright stars packed loosely into a circle with a diameter 50 percent larger than the Full Moon.

IC 4665's large size arises in part because we view it from about 1,100 light-years away. This young open cluster formed only some 40 million years ago. Although most youthful clusters lie close to the Milky Way's plane, IC 4665 appears 17° away. Like globular cluster M5, this open cluster doesn't receive a lot of traffic. Observers tend to overlook it because IC objects tend to be faint and rather unexciting.

IC 4665
Constellation:
Ophiuchus the Serpent-bearer
Right ascension:
17h46m
Declination: 5°39'
Magnitude: 4.2
Apparent size: 45'
Distance: 1,100 light-years

Tour three summer globulars in Ophiuchus

Some 200,000 stars call globular M12 home, but it likely held 1 million low-mass stars that were ripped away as the cluster passed through the Milky Way's disk. ESA/HUBBLE & NASA

Break out your star charts to track down the sprawling constellation Ophiuchus the Serpent-bearer, which resides north of Scorpius and south of Hercules, two more prominent constellations. Binoculars will show you three nice globular star clusters.

Start at the westernmost bright star in Ophiuchus, magnitude 2.7 Delta (δ) Ophiuchi. With your binoculars, head 11° due east and you should sweep up M10 (NGC 6254). M12 (NGC 6218) lies 3.3° northwest of M10, so you'll see the two clusters in the same field of view. To find M9

(NGC 6333), move south-southeast about two binocular fields to magnitude 2.4 Eta (η) Ophiuchi. M9 lies 3.5° southeast of this star.

All three globulars make easy binocular targets and show up well with direct vision through an 80mm wide-field telescope. M10 and M12 are both resolved at their outer edges with the 80mm scope at 75x or higher powers. Because M9 lies nearly twice as far away as the other two, it only

hints at resolution by appearing grainy.

While you explore Ophiuchus, you might want to track down a couple more nice globular clusters. M107 (NGC 6171) lies 2.7° south-southwest of magnitude 2.6 Zeta (ζ) Ophiuchi and M14 (NGC 6402) stands 10° due east of M10. Both of these 8th-magnitude Messier clusters show up through binoculars and are easy targets with an 80mm scope.

The 7th-magnitude globular M10 is the brightest of our three objects. The cluster spans about two-thirds the diameter of the Full Moon. N.A. SHARP, VANESSA HARVEY/REU PROGRAM/NOIRLAB/NSF/AURA

The Hubble Space Telescope reveals a range of colors for the stars in globular cluster M9 (red means cool, blue hot) and resolves stars to the core in this image. NASA/ESA

M12
Also known as:
NGC 6218
Constellation:
Ophiuchus the Serpent-bearer
Right ascension:
16h47m
Declination: –1°57'
Magnitude: 6.7
Apparent size: 16'
Distance: 16,000 light-years

M10
Also known as:
NGC 6254
Constellation:
Ophiuchus the Serpent-bearer
Right ascension:
16h57m
Declination: –4°06'
Magnitude: 6.6
Apparent size: 20'
Distance: 14,000 light-years

M9
Also known as:
NGC 6333
Constellation:
Ophiuchus the Serpent-bearer
Right ascension:
17h19m
Declination: –18°31'
Magnitude: 7.7
Apparent size: 9.3'
Distance: 26,000 light-years

Globular clusters in Hercules

Globular clusters tend
to concentrate toward the
Milky Way's center and
summer is when Earth
faces the galactic core at
night, so summer means
globular season. Let's tour
two of the best summer
globulars and one less-
visited cluster in Hercules
the Strongman.

We start this journey
with the best of the three.
The Hercules Cluster (M13)
shines at magnitude 5.8
and shows up to the naked
eye from a dark-sky site.
And it's easy to find

because it lies on the western edge of the Strongman's
conspicuous Keystone
asterism. Draw an imagi-
nary line from magnitude
3.5 Eta (η) Herculis at the
Keystone's northwestern
corner to magnitude 2.8
Zeta (ζ) Herculis at the
southwestern corner. Aim
your binoculars one-third
of the way along this line
and you'll see M13 as a
fuzzy ball. An easy target
with an 80mm telescope,
the cluster appears at the
verge of resolution when
viewed at low power. You

can start to resolve M13's
stars at the cluster's edge
with 50x to 100x.

To find M92 (NGC
6341), go to the eastern side
of the Keystone and center
your binoculars on the
northern star, magnitude
3.1 Pi (π) Herculis. Then
slide 6.3° due north and
you should sweep up M92.
This magnitude 6.4 cluster
shows up easily through
binoculars and 80mm
scopes. Although M92 lies
about the same distance
from Earth as M13, it holds
only one-third the mass.

**Most observers consider
the Hercules Cluster (M13)
the finest globular in the
northern sky, and even a
quick look through a
telescope will show you
why.** T.A. RECTOR (UNIVERSITY OF ALASKA
ANCHORAGE) AND H. SCHWEIKER (WIYN AND
NOIRLAB/NSF/AURA)

This makes it look noticeably less condensed.

Our last target,
NGC 6229, resides 4.8°
east-northeast of magni-
tude 3.9 Tau (τ) Herculis.
NGC 6229 lies nearly
100,000 light-years from
Earth and appears about 25
times dimmer than M13.
Still, it's a decent globular,
though nothing compared
with Hercules' other two.

The Hercules Cluster

Also known as: M13, NGC 6205

Constellation: Hercules the Strongman

Right ascension: 16h42m

Declination: 36°28'

Magnitude: 5.8

Apparent size: 20'

Distance: 23,000 light-years

Globular cluster M92 (NGC 6341) lies just 10° northeast of M13 and shows up as a fuzzy haze through binoculars. NEIL FLEMING

M92

Also known as: NGC 6341

Constellation: Hercules the Strongman

Right ascension: 17h17m

Declination: 43°08'

Magnitude: 6.4

Apparent size: 14'

Distance: 27,000 light-years

NGC 6229

Constellation: Hercules the Strongman

Right ascension: 16h47m

Declination: 47°32'

Magnitude: 9.4

Apparent size: 4.5'

Distance: 99,000 light-years

NGC 6229 resides in northern Hercules. Although it pales in comparison with this constellation's two Messier globulars, it's still worth viewing through a telescope. MARTIN GERMANO

The False Comet and its neighborhood

The area just north of Zeta Scorpii (the pair of stars at bottom) sparkles with bright clusters and nebulae. The bright cluster just above Zeta is NGC 6231 while the Prawn Nebula glows red near the image's top. KPNO/NOIRLAB/NSF/AURA/ADAM BLOCK

NGC 6231 ranks as the sixth-brightest open star cluster in the entire sky. Some observers refer to it as the Northern Jewel Box. KPNO/NOIRLAB/NSF/AURA/ADAM BLOCK

Swing south to the constellation Scorpius the Scorpion and you'll find an unusual collection of stars, star clusters, and emission nebulae, with the whole assortment having an evocative name: the False Comet.

The structure lies in the Scorpion's tail, where it begins to curl from south to east. The corner star actually comprises a stellar pair—Zeta[1] (ζ^1) and Zeta[2] (ζ^2) Scorpii—that show up to the unaided eye under a dark sky if you observe from far enough south. Just 0.5° north of these two lies the star cluster NGC 6231. To the eye, it looks a little like the head of a comet, and running north from there you'll find two more

Northern Jewel Box
Also known as: NGC 6231, Caldwell 76
Constellation: Scorpius the Scorpion
Right ascension: 16h54m
Declination: –41°50'
Magnitude: 2.6
Apparent size: 14'
Distance: 4,100 light-years

Collinder 316
Constellation: Scorpius the Scorpion
Right ascension: 16h55m
Declination: –40°50'
Magnitude: 6.4
Apparent size: 60'
Distance: 3,700 light-years

This close-up of the Prawn Nebula's northeastern section is chock full of young stars and the glowing clouds of gas and dust that give birth to them. ESO

star clusters and an emission nebula that fan out to form the False Comet's tail.

NGC 6231 may be small but it packs a punch: It ranks as the sixth-brightest open star cluster in the sky. Through an 80mm refractor, it showcases eight bright stars and about 50 dimmer ones. The brightest of these radiate 60,000 times the light of the Sun. All told, this small cluster shines brighter than the multimillion-star globular star cluster Omega Centauri. Lots of stars

reside north of NGC 6231 in what astronomers have determined to be at least two actual groups of stars: Collinder 316 and Trumpler 24. To the naked eye, this whole region appears as one sprawling cluster larger than the Pleiades (M45) in Taurus. All these clusters belong to a huge grouping of energetic stars called the Scorpius OB1 association.

At the northeastern end of Trumpler 24 lies the Prawn Nebula (IC 4628). Binoculars give a hint of

this emission region while an 80mm telescope shows it even without a nebular filter. An OIII filter helps bring out the nebula, also cataloged as Gum 56, as a ribbon running mostly east to west.

As a whole, the area from Zeta to IC 4628 spans about 2.5°, or five times the Full Moon's size. It's an amazing vista best appreciated with a small wide-angle telescope. Sharp-eyed observers may also detect the dark nebula Barnard 48 just east of the Prawn.

The Prawn Nebula

Also known as: IC 4628, Gum 56

Constellation: Scorpius the Scorpion

Right ascension: 16h57m

Declination: –40°31'

Magnitude: —

Apparent size: 90'x60'

Distance: 6,000 light-years

The Pipe and Snake nebulae

To see the Pipe Nebula (Barnard 59, 65-7, 78), remember to look for something exceedingly large—it spans 8°. Your best view will come with the naked eye or wide-angle binoculars. To find it, first locate 1st-magnitude Antares, the brightest star in Scorpius, and 3rd-magnitude Lambda (λ) Sagittarii, the star marking the lid of Sagittarius' Teapot asterism. Now aim your binoculars midway between these stars and you should see a dark pattern that looks like a tobacco pipe. The stem runs east to west while the bowl sits on the east side pointing into Ophiuchus. You need a dark sky to see the shape because it appears silhouetted against a background of faint stars.

The wide-field Milky Way shot below easily shows the Pipe Nebula in the upper half of the image starting in the middle and ending about halfway to the left side. The structure is too large for any telescope, although a wide-field 80mm will let you scan the entire object with minimal movement. The best views come through 8x42 or 10x50 binoculars.

To the northwest of the Pipe's bowl lies a snake in the grass of the starry Milky Way. The Snake Nebula (Barnard 72) looks like a curl of smoke climbing out of the Pipe. It measures only about 6' long, so it's much smaller than the Pipe. Like the Pipe, the Snake is a dark nebula that shows up in silhouette against the Milky Way. A wide-field 80mm telescope at low power makes an excellent choice for examining the Pipe's intricate details.

If you have access to a larger telescope, you will find two other deep-sky objects that bracket the Snake Nebula. The Little Ghost Nebula (NGC 6369) is a faint planetary nebula 1.3° to the east and the faint globular cluster NGC 6325 stands 1.3° to the west. You'll need at least a 6-inch telescope to spot these objects.

The Pipe Nebula

Also known as: Barnard 59, 65–7, 78
Constellation: Ophiuchus the Serpent-bearer
Right ascension: 17h30m
Declination: –25°00'
Magnitude: —
Apparent size: 500'x140'
Distance: 650 light-years

The Snake Nebula

Also known as: Barnard 72
Constellation: Ophiuchus the Serpent-bearer
Right ascension: 17h24m
Declination: –23°42'
Magnitude: —
Apparent size: 6'
Distance: 650 light-years

This wide-field view of the Milky Way in Ophiuchus shows the Pipe Nebula starting in the frame's middle and ending about halfway to the left side. The dark nebula appears to the naked eye under a dark sky. ESO/S. GUISARD

This close-up of the Pipe Nebula reveals the Pipe's stem on the right-hand side and the bowl extending to the upper left. ESO/Y. BELETSKY

The dark Snake Nebula slithers across the top of this image, almost as if it is smoke wafting from the Pipe Nebula. GEOFF SMITH

The Butterfly Cluster

The Butterfly Cluster (M6) stands out north of the Scorpion's stinger. While viewing the cluster, try to detect the orange color of the bright variable star BM Scorpii. CHRIS SCHUR

One of the finest sights under a dark summer sky awaits viewers within the confines of Scorpius the Scorpion. First, find the two stars marking the Scorpion's stinger: magnitude 1.6 Shaula (Lambda [λ] Scorpii) and magnitude 2.7 Lesath (Upsilon [υ] Scorpii). Then scan about 5° northeast and you will see two fuzzy areas with your naked eye. The larger one lies farther east, and we'll cover it on the next page. The northern one is the magnificent Butterfly Cluster (M6).

This open cluster glows at magnitude 4.2 and makes a fine sight with the naked eye, binoculars, or an 80mm telescope. The Butterfly holds about 80 stars in a condensed clump about 0.4° across (slightly smaller than the Full Moon). The brightest one, BM Scorpii, is a giant star with a distinct orange hue, and it varies in brightness from 6th to 9th magnitude with a period of more than two years. The Butterfly got its name because it appears pinched along its middle. Italian astronomer Giovanni Battista Hodierna gets credit for first reporting the cluster, though as a naked-eye object, other observers undoubtedly spotted it earlier.

Astronomers estimate M6 formed nearly 100 million years ago. It resides in a complex star field close to the plane of the Milky Way and less than 4° from the center of our galaxy. As you might suspect, several other treats await a survey of the cluster's vicinity through a wide-field telescope. Nearly in contact with M6 to the east stands the dark nebula Barnard 278; a similar dark nebula, Barnard 275, lies to the west. Both appear roughly circular. About 0.5° west of M6 lies the edge of the emission patch RCW 132. This irregularly shaped nebula spans roughly 2° and contains the star cluster NGC 6383. An OIII filter enhances the emission.

The Butterfly Cluster

Also known as: M6, NGC 6405

Constellation: Scorpius the Scorpion

Right ascension: 17h40m

Declination: –32°15'

Magnitude: 4.2

Apparent size: 25'

Distance: 1,600 light-years

Ptolemy's Cluster

Although the Butterfly Cluster would dominate most regions, it doesn't even rule the area northeast of the Scorpion's stinger. That title belongs to Ptolemy's Cluster (M7), which appears twice as bright and more than three times the diameter of its neighbor. M7 glows at magnitude 3.3 and spans 80'—well over twice the Full Moon's diameter. It's easy to spot with the naked eye and spectacular through binoculars and small telescopes.

Ptolemy's Cluster ranks among the most beautiful open star clusters because it appears bright, rich in luminous stars, and symmetrical. At low power, you can see the cluster in contrast with the comparatively empty sky; at higher powers, you can tease out some background objects. Viewing with anything larger than a 6-inch scope may show intriguing features buried within the cluster, but you'll lose some of the aesthetic appeal.

After you appreciate the cluster itself, hunt down a few of its neighbors. The dark nebula Barnard 283 touches M7's northwestern edge while Barnard 286 lies about 0.5° south of the cluster and Barnard 287 finds itself embedded along M7's southern periphery.

Globular cluster NGC 6453 overlaps M7's western edge, but at 10th magnitude, it is a challenge for an 80mm scope. A bit farther afield, find magnitude 3.2 G Scorpii 2.4° south-southwest of M7. Look at this star with a telescope and try to spot the globular cluster NGC 6441 just to the east. Although difficult to see so close to the star's glare, the object is one of the easiest-to-find globulars below naked-eye visibility.

Ptolemy's Cluster
Also known as: M7, NGC 6475
Constellation: Scorpius the Scorpion
Right ascension: 17h54m
Declination: –34°49'
Magnitude: 3.3
Apparent size: 80'
Distance: 980 light-years

Ptolemy's Cluster (M7) in Scorpius shines big and bright near the eastern edge of Scorpius the Scorpion. ESO

The Milky Way's heart

Under a dark sky, look at the Teapot asterism in Sagittarius the Archer. Some 6° north of the Teapot's spout you'll see the soft glow of an emission nebula. Binoculars reveal two patches of nebulosity in the same field. The larger, southern one is the Lagoon Nebula (M8) and the northern one is the Trifid Nebula (M20). With a small wide-field telescope, this area ranks among the best vistas in the sky.

The Lagoon is a glorious sight that a small telescope will resolve into an open star cluster, one of the brightest emission patches in the sky, and associated dark clouds. Look carefully at the center of the nebula and you should start to detect a large dust lane (the "lagoon") that bifurcates

The pinkish hues of the Lagoon Nebula (M8) stand out on the right side of this image while the bluish tones of the Trifid Nebula (M20) grace the upper right corner. ESO/DSS2

The Lagoon Nebula
Also known as: M8, NGC 6523
Constellation: Sagittarius the Archer
Right ascension: 18h04m
Declination: –24°23'
Magnitude: 6.0
Apparent size: 90'x40'
Distance: 4,100 light-years

The Trifid Nebula
Also known as: M20, NGC 6514
Constellation: Sagittarius the Archer
Right ascension: 18h03m
Declination: –22°58'
Magnitude: 6.3
Apparent size: 28'
Distance: 4,000 light-years

the emission nebula. With an east-to-west extent of 90′, M8 spans three times the Full Moon's diameter.

The Trifid is an interesting beast. Its southern half is an emission nebula with a multiple star system at its core. Three irregular dust lanes trisect this emission patch and give the nebula its name. These may be tough to see in a small wide-field telescope; try higher power or an OIII filter to boost the contrast. The nebula's northern portion seems fainter, and it is. It's a reflection nebula, a dust cloud illuminated by the light of the bright star at its center.

With a wide-field telescope, several more objects in the immediate vicinity are worth tracking down. Less than 1° northeast of M20 lies the nice open star cluster M21 (NGC 6531). Two globular clusters lie southeast of the Lagoon: Magnitude 7.8 NGC 6544 stands 1° from M8's core and magnitude 8.1 NGC 6553 lies 1° farther on.

If you want to stretch your observing skills, try to find the faint emission patch NGC 6526 almost directly between M8 and M20. It is about 70 percent M8's size. Some 2° east and slightly north of M8 stands another large emission patch, NGC 6559, which is intertwined with fainter IC 4685, IC 1275, and IC 1274.

The Trifid Nebula (M20) consists of both emission (pink) and reflection (blue) nebulosity. Its common name comes from the three dark dust lanes silhouetted in front. ESO

The Lagoon Nebula (M8) ranks among the brightest emission nebulae in the sky. The dark rift that cuts north to south through its center represents the lagoon. ESO/VPHAS+ TEAM

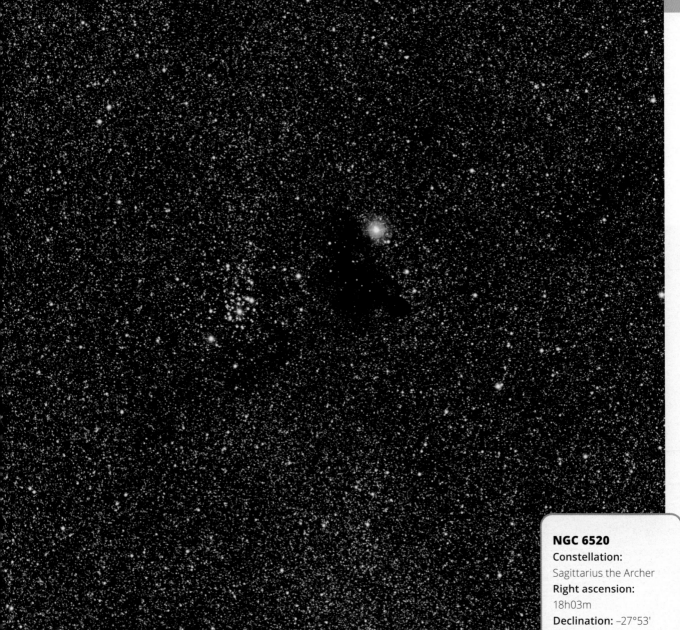

Open cluster NGC 6520 and the Ink Spot

Starting at the Lagoon Nebula, slew your telescope 3.5° due south. Although you'll be looking into the brightest portion of the Milky Way, the Large Sagittarius Star Cloud, a small group of stars should catch your attention. The tightly packed open cluster NGC 6520 glows at 8th magnitude against the rich star field. Binoculars give a surprisingly nice view.

But the most intriguing object in this area lies just west of the cluster. Increase your telescope's magnification and you should spot a tiny, irregularly shaped patch of inky-black sky. This opaque cloud—the Ink Spot (Barnard 86)—makes a superb contrast to the cluster next door. It lies beyond the cluster, but in front of the Milky Way's myriad stars.

The Ink Spot (Barnard 86) appears as a black cloud set against countless background stars and the neighboring cluster NGC 6520 in the brightest section of the Milky Way. ESO

NGC 6520
Constellation:
Sagittarius the Archer
Right ascension:
18h03m
Declination: –27°53'
Magnitude: 7.6
Apparent size: 5'
Distance: 5,100
light-years

The Ink Spot
Also known as:
Barnard 86
Constellation:
Sagittarius the Archer
Right ascension:
18h03m
Declination: –27°52'
Magnitude: —
Apparent size: 5'
Distance: 6,000
light-years

Globular cluster NGC 6541 and the Emerald Nebula

South of the Teapot asterism in Sagittarius lies the small constellation Corona Australis the Southern Crown. Despite its diminutive size, Corona Australis harbors the fine globular star cluster NGC 6541 (Caldwell 78), which glows brightly at magnitude 6.3. Astronomers think the stars in NGC 6541 formed at about the same time as our galaxy, some 13 billion years ago.

If you can see the semicircular pattern of stars that forms the Crown's shape, follow the arc south and then west to a crooked line of 5th- and 6th-magnitude stars that stretches about 5° and points to NGC 6541. Binoculars or a wide-field telescope will let you sweep up the cluster with little difficulty.

Our next target resides about 50° due north of the globular, in the northeastern reaches of Ophiuchus the Serpent-bearer. The remarkable Emerald Nebula (NGC 6572) lies 7.5° east-northeast of 3rd-magnitude Beta (β) Ophiuchi. This planetary nebula got its common name because it sports a remarkable greenish color visible through almost any telescope.

Finding NGC 6572 with a small wide-field scope proves a challenge not because it is faint (magnitude 8.1) but because it is tiny (just 14" across). Although it appears stellar at low power, raise the magnification to between 50x and 100x and you should see it as more than a point of light. The real key to identifying it, however, is its greenish color. The color's intensity may surprise you, and it shows up best if you magnify the nebula enough to see a small disk.

NGC 6541
Also known as:
Caldwell 78
Constellation: Corona Australis the Southern Crown
Right ascension: 18h08m
Declination: −43°43'
Magnitude: 6.3
Apparent size: 13'
Distance: 24,000 light-years

The Emerald Nebula
Also known as:
NGC 6572
Constellation: Ophiuchus the Serpent-bearer
Right ascension: 18h12m
Declination: 6°51'
Magnitude: 8.1
Apparent size: 14"
Distance: 3,500 light-years

Globular cluster NGC 6541 ranks as the finest deep-sky object in Corona Australis. The cluster looks great through small telescopes, but don't expect to see the color and detail the Hubble Space Telescope captured. NASA/ESA/G. PIOTTO (UNIVERSITÀ DEGLI STUDI DI PADOVA)

The greenish hue of the tiny Emerald Nebula (NGC 6572) stands out through most telescopes. Like all planetary nebulae, NGC 6572 represents the death throes of a Sun-like star. ADAM BLOCK

The Small Sagittarius Star Cloud and M18

The Small Sagittarius Star Cloud (M24) stands out to the naked eye as a bright patch in the summer Milky Way. Take time to explore the cloud with binoculars or a wide-field telescope.

CHRIS SCHUR

If you head back to the spout of the Teapot asterism in Sagittarius and look about 10° to the north, you'll see a bright oval patch of the Milky Way that trends from northeast to southwest. This is the Small Sagittarius Star Cloud (M24)—a must-see object in the summer sky. The star cloud appears obvious to the naked eye once you spend some time examining the brighter portions of the Milky Way in this area. It stands out well in binoculars and proves to be an easy target through an 80mm wide-field telescope.

The richness of the star background in this area is amazing if you use a wide-angle eyepiece to simply wander around. The

The Small Sagittarius Star Cloud

Also known as: M24
Constellation: Sagittarius the Archer
Right ascension: 18h17m
Declination: –18°31'
Magnitude: 4.5
Apparent size: 2°×1°
Distance: 10,000 light-years

The magnitude 7.5 glow of open cluster M18 (NGC 6613) comes from a smattering of roughly three dozen faint stars. ANTHONY AYIOMAMITIS

name "star cloud" is a bit misleading. M24 is not a specific deep-sky object. Instead, when we look in this direction, we're gazing through a window in a nearby spiral arm into denser regions of our galaxy more than halfway to the core.

If you want more objects to tackle, try looking in the northeastern portion of the cloud for NGC 6603, a condensed open star cluster. Then shift your gaze to the cloud's northwestern side to find two relatively small dark nebulae— Barnard 92 and Barnard 93— silhouetted against the star cloud. Barnard 92 stands out more noticeably. Finally, if you are observing with a 6-inch or larger telescope, look for the 11th-magnitude planetary

nebula NGC 6567 in M24's southwestern sector. This target is too faint and small to see easily through smaller scopes.

After you've had your fill gazing at the brighter portions of M24, move about 1° north from the star cloud's eastern section. You'll come upon a small, loose, open cluster of about 40 faint stars, which made it as the 18th entry in the Messier Catalog. If you look carefully at M18's brightest stars, you'll see a bit of nebulosity. If you go a little farther north of M18, you'll run into the wonderful Omega Nebula (M17), but that's a story for the next two pages.

M18
Also known as:
NGC 6613
Constellation:
Sagittarius the Archer
Right ascension:
18h20m
Declination: –17°06'
Magnitude: 7.5
Apparent size: 9'
Distance: 4,000
light-years

Star formation in Sagittarius and Serpens

This broad swath of the Milky Way contains three outstanding nebulae: Sharpless 2-54 (top), the Eagle Nebula (M16, middle), and the Omega Nebula (M17, bottom). ESO

Summer is the time of year to view one of the most complex and intricate vistas in the entire Milky Way. This grouping of three nebulae stretches from northern Sagittarius the Archer into the southern part of Serpens Cauda the Serpent's Tail.

Let's tour this area with binoculars first. I would start at the Lagoon Nebula (M8) because it's large and shines brightly enough to see easily. Once you find it, slide northward along the Milky Way about 7° until you reach the Small Sagittarius Star Cloud (M24). Continue north for 1.4° and you will detect the small but fairly bright star cluster M18, and in another 0.9° you'll come to the first of our three nebulae: the Omega Nebula or Swan Nebula (M17).

Put M17 at the bottom of your binocular field and you should see another bright spot near the middle of the field. This is the Eagle Nebula (M16), immortalized by the Hubble Space Telescope in its "Pillars of Creation" image. Toward the top of the binocular field, about half the distance as between M17 and M16, you should see another large area of fainter nebulosity. The third of our three nebulae surrounds the open star cluster NGC 6604; the nebulosity itself is designated Sharpless 2-54.

To duplicate the three nebulae in a single field through a telescope, you'll need an ultra-wide field of about 5°. You can accomplish this using a wide-field 80mm scope with a 30mm 82° apparent field of view or wider, but this is a fairly rare eyepiece. Let's stick with the "usual" 22mm 68° apparent field of view eyepiece, which yields a bit under a 3° field on a wide-field 80mm scope. By moving the telescope one field of view south to north, you can cover from the star cluster M18 just south of M17 to the top of Sharpless 2-54 at the north. Move the field twice and you can scan from the bottom of the Small Sagittarius Star Cloud to the top of Sharpless 2-54, a stretch of 9° along the summer Milky Way in an impressive and complex region. Far more objects exist in this wide swath than we can cover in this book, so I'll briefly describe the appearance of the three primary nebulae and mention some of the other targets in the area you can track down.

At 23 power, the Omega Nebula's main "bar" as well as the fainter outer gas clouds show up easily. Higher power reveals more details in the region. Although you do not need to use an OIII filter, one helps you to see M17's outer extensions. If you look 1° south of M17, you'll find the small bright open cluster M18. A wide-field 80mm scope resolves it at higher powers, and if you look carefully at its brighter stars, you should detect traces of nebulosity.

Low powers show the Eagle Nebula (M16) as well as its inner star cluster, NGC 6611. The hot young stars within this cluster

The Eagle Nebula (M16) is an area of glowing gas highlighted by the dark body and wings of a raptor (center) the Hubble Space Telescope made famous as the Pillars of Creation. ESO

The faint reddish tendrils of Sharpless 2-54 wrap around the bottom of the bright star cluster NGC 6604 at upper left. ESO

emit the ultraviolet radiation that ultimately drives the Eagle's emission. At higher powers, you can start to discern signs of mottling delineating the Pillars of Creation, but you really need a 12- to 14-inch scope to see them well.

While small scopes mostly resolve the star cluster NGC 6604, the Sharpless 2-54 nebula shows up best either through binoculars or when using an OIII filter. The nebula appears most distinct to the west of the cluster.

Stunning details emerge in this wide-angle view of the Omega Nebula (M17) in Sagittarius. ESO

The scattered stars of the Tweedledum Cluster (NGC 6633) shine brightly enough to see with the naked eye near the edge of the Milky Way in northeastern Ophiuchus. ANTHONY AYIOMAMITIS

The Tweedledee and Tweedledum clusters

A pair of pretty open star clusters greets observers who can tear themselves away from the wonders of Sagittarius. Shift your gaze to northeastern Ophiuchus the Serpent-bearer to pick up the Tweedledum Cluster (NGC 6633) and then head 3.1° east-southeast into northern Serpens to find its companion, the Tweedledee Cluster (IC 4756). NGC 6633 appears as a fuzzy patch of light to the naked eye and shows up easily through binoculars. Look for it by scanning 11° (nearly two binocular fields) east and slightly north of 3rd-magnitude Beta (β) Ophiuchi. Both Tweedledum and Tweedledee fit into a single binocular field, and both make nice targets through an 80mm telescope. The pair resides against a rich Milky Way background.

A telescope reveals NGC 6633 as a magnitude 4.6 open star cluster nearly the size of the Full Moon. The loosely packed grouping shows about 30 stars at low power; higher powers double that total. With no obvious central condensation, the cluster looks best at low power. Astronomers estimate that NGC 6633 lies 1,300 light-years from Earth and formed roughly 660 million years ago.

IC 4756 shines as brightly as NGC 6633 but appears about 50 percent larger than its neighbor.

That's why most observers consider IC 4756 to be dimmer. Like Tweedledum, Tweedledee has no strong central condensation and thus looks best through binoculars or a wide-field telescope. IC 4756 holds about 50 stars and shines across a distance of 1,600 light-years.

The Tweedledee Cluster

Also known as: IC 4756, Graff's Cluster
Constellation: Serpens the Serpent
Right ascension: 18h39m
Declination: 5°28'
Magnitude: 4.6
Apparent size: 39'
Distance: 1,600 light-years

The Tweedledum Cluster

Also known as: NGC 6633
Constellation: Ophiuchus the Serpent-bearer
Right ascension: 18h27m
Declination: 6°34'
Magnitude: 4.6
Apparent size: 27'
Distance: 1,300 light-years

The Tweedledee Cluster (IC 4756) lies in Serpens just 3.1° from NGC 6633. Tweedledee glows as brightly as its neighbor but covers twice the area. ROBERTO MURA/CC BY-SA 3.0

The Wild Duck Cluster

The 5th-magnitude Wild Duck Cluster (M11) lies in northeastern Scutum. This tightly packed group boasts 100 or so stars visible through a small telescope. ESO

The tiny constellation
Scutum the Shield harbors one of the richest open star clusters in the sky: the Wild Duck Cluster (M11). M11 lies in the much larger Scutum Star Cloud, a bright knot in the Milky Way north of the Small Sagittarius Star Cloud and south along the Milky Way from 1st-magnitude Altair in Aquila the Eagle.

To find the cluster, first locate Altair—the southernmost star of the Summer Triangle. From there, trace the Eagle's pattern south to its tail feathers, 3rd-magnitude Lambda (λ) and 4th-magnitude 12 Aquilae. Then pick up 5th-magnitude Eta (η) Scuti across the border in Scutum. These three stars form a westward-curving arc that points to M11 just 1.5° farther west.

You may glimpse 5th-magnitude M11 with your naked eye under a dark sky. Binoculars show the bright cluster easily. A small telescope reveals a beautiful object with more than 100 uniformly bright members packed in a small area. (The cluster's total population runs into the thousands.) M11's stars form a vague triangular shape that reminded some early observers of a flock of ducks in flight.

Through an 80mm wide-field telescope, M11 ranks among the most memorable sights in the summer sky. At higher powers, observers start to see knots and clumps of stars along with a few dark lanes. The area surrounding M11 includes a few other objects of interest, including dark dust clouds just to its north and south and the 8th-magnitude globular cluster NGC 6712 some 2.5° to the south.

> **The Wild Duck Cluster**
> **Also known as:** M11, NGC 6705
> **Constellation:** Scutum the Shield
> **Right ascension:** 18h51m
> **Declination:** –6°16'
> **Magnitude:** 5.3
> **Apparent size:** 14'
> **Distance:** 6,100 light-years

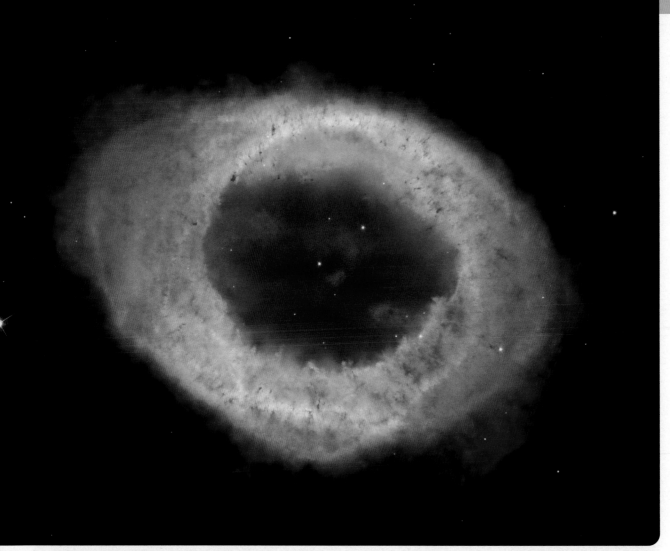

The Ring Nebula

Most emission nebulae are vast clouds of gas and dust collapsing under the force of gravity to create new stars. The hot young stars forming within excite the surrounding gas and cause it to glow. Plenty of these stellar maternity wards dot the summer sky. But some emission regions mark sites at the other end of a star's life. A planetary nebula represents a dying star gently puffing off its outer layers. The white dwarf star left behind burns hot and energizes the expanding gas to luminescence.

The Ring Nebula (M57) in Lyra the Harp epitomizes this stage in a Sun-like star's life. It makes a prime example not only because it shines brightly but also because it's easy to find. First locate magnitude 0.0 Vega—the brightest and westernmost star of the Summer Triangle. Then look southeast for the relatively small parallelogram of conspicuous stars that forms the rest of the Harp's shape. M57 lies midway between the parallelogram's southernmost stars, Gamma (γ) and Beta (β) Lyrae.

It takes only a small telescope to spot this planetary. In a wide-field 80mm scope at low power, M57 looks like a small, colorless smoke ring. Boost the power to make sure you see a brighter outer ring. To see M57's central white dwarf, you need a 12-inch or larger telescope and a high-power eyepiece to tease it from the background.

The Ring Nebula lies about 2,600 light-years from Earth and spans about 1 light-year. The ring itself grows about 1" per century as the expelled gas rushes away from the central star at some 55,000 mph. Like all planetary nebulae, the Ring eventually will expand so much that it will dissipate and no longer be visible.

The Ring Nebula
Also known as: M57, NGC 6720
Constellation: Lyra the Harp
Right ascension: 18h54m
Declination: 33°02'
Magnitude: 8.8
Apparent size: 1.4'x1'
Distance: 2,600 light-years

Star clusters NGC 6755 and NGC 6756

For a constellation located along the spine of the Milky Way, Aquila the Eagle has few bright deep-sky objects. Not even 18th-century French astronomer Charles Messier deemed to include any of the Eagle's deep-sky objects in his respected catalog. One of its best lies in eastern Aquila, not far from its border with Serpens. First zero in on 3rd-magnitude Delta (δ) Aquilae at the base of the Eagle's wings, then scan 4.6° west-northwest with your binoculars. There you'll find the magnitude 7.5 open star cluster NGC 6755.

Through a wide-field 80mm telescope, NGC 6755 proves to be an attractive cluster of about 100 stars measuring 15' across (half the Full Moon's diameter). Astronomers estimate this object lies 8,100 light-years away and formed 250 million years ago.

NGC 6755 even appears to have a friend. Open cluster NGC 6756 lies 0.5° north-northeast of its neighbor. Unfortunately, NGC 6756 glows dimly at magnitude 10.6 and poses a real challenge for observers using an 80mmm scope.

A 6-inch instrument delivers much better views of this cluster pair. NGC 6756 resides about 6,400 light-years from Earth and started life barely 60 million years ago. The differences in distance and age between the two clusters can mean only one thing: They are unrelated and a mere chance alignment on the sky.

NGC 6755 (lower right) and NGC 6756 to its upper left form a nice double cluster through small- and medium-sized telescopes. The bigger and brighter NGC 6755 stands out better from the Milky Way background. BERNHARD HUBL

NGC 6755
Constellation: Aquila the Eagle
Right ascension: 19h08m
Declination: 4°16'
Magnitude: 7.5
Apparent size: 15'
Distance: 8,100 light-years

NGC 6756
Constellation: Aquila the Eagle
Right ascension: 19h09m
Declination: 4°42'
Magnitude: 10.6
Apparent size: 4'
Distance: 6,400 light-years

Brocchi's Cluster comprises 10 stars: six that form a straight line running east-to-west and four that make an arc curving southward. It gives the appearance of an upside-down coathanger. ANTHONY AYIOMAMITIS

Brocchi's Cluster

Few constellations look as uninspiring as Vulpecula the Fox. Its brightest star, Alpha (α) Vulpeculae, glows at magnitude 4.4 and doesn't show up to the naked eye under even moderately light-polluted skies. You can spy a few of its brighter stars under a dark sky if you use the more conspicuous pattern of Sagitta the Arrow just to its south as a guide.

Despite its faintness, Vulpecula does hold a few deep-sky objects worth your time. Perhaps the most intriguing lies in the constellation's southwestern corner. Scan with binoculars 4.5° due south of Alpha Vulpeculae and you'll come to Brocchi's Cluster. Named after American amateur astronomer Dalmero Francis Brocchi, the cluster more often goes by the name "the Coathanger." The reason becomes obvious when you view the object through binoculars or a wide-field telescope.

Although a few of its stars show up to the naked eye, you'll need some optical aid to make out the Coathanger's figure. Six stars form the shape's straight-line base while another four suns curve away to the south to outline the hanger. All glow between 5th and 7th magnitude. Most of the stars in Brocchi's Cluster appear white, but one of them in the handle has a noticeable orange cast.

Despite its name, Brocchi's Cluster is a chance alignment of stars forming an asterism, or recognizable pattern of stars. Measurements show the members range from a bit more than 200 light-years to more than 1,000 light-years away.

Brocchi's Cluster
Also known as:
The Coathanger,
Collinder 399
Constellation: Vulpecula
the Fox
Right ascension:
19h25m
Declination: 20°11'
Magnitude: 3.6
Apparent size: 60'

The Dumbbell Nebula

The Dumbbell Nebula (M27) in Vulpecula stands out among the sky's planetary nebulae. Even small scopes reveal the main structure's pinched appearance and slightly greenish hue. ESO/I. APPENZELLER/W. SEIFERT/O. STAHL/M. ZAMANI

Vulpecula's bigger claim to deep-sky fame resides in the Fox's south-central sector. The Dumbbell Nebula (M27) appears as a small, round, misty spot through 10x50 binoculars. To locate it, start at magnitude 3.5 Gamma (γ) Sagittae—the easternmost star in Sagitta's arrow pattern—and travel 3.2° due north. The Dumbbell is a planetary nebula, the final stage in the life of a Sun-like star.

M27 glows at magnitude 7.4, bright enough to show up through binoculars or a finder scope. But it truly shines when viewed through a telescope. You can magnify this planetary quite a bit even with an 80mm refractor. Expect to see a relatively large, bright ball of glowing gas somewhat pinched along its east-west axis. Larger scopes reveal more patchiness and perhaps a trace of green color. The Dumbbell ranks as the best planetary nebula both in this book and in the sky—with only the Ring Nebula (M57) in Lyra serious competition. This is one object you don't want to miss as you explore the summer sky.

The Dumbbell Nebula

Also known as: M27, NGC 6853
Constellation: Vulpecula the Fox
Right ascension: 20h00m
Declination: 22°43'
Magnitude: 7.4
Apparent size: 8'x6'
Distance: 1,300 light-years

Barnard's E

E.E. Barnard discovered the dark dust lanes in Aquila whose silhouette against the bright background of the Milky Way forms a giant letter E. GEORGE GREANEY

Aquila the Eagle rides high in the evening sky in late summer. Although the Milky Way and several bright stars call this constellation home, I want to focus now on an intriguing dark area. Start at Aquila's luminary, 1st-magnitude Altair (Alpha [α] Aquilae) and then travel 2.1° north-northwest to magnitude 2.7 Tarazed (Gamma [γ] Aquilae). Now slide 1.6° (about one-quarter of a binocular field) to the west and look for a series of dark dust lanes slightly larger than the Full Moon. The dust forms a giant letter E, which typically shows up better with averted vision.

Barnard's E comprises the U-shaped dark nebula Barnard 143 and the more-or-less straight dark nebula Barnard 142 arranged to create the appearance of our alphabet's fifth letter. Barnard 143 appears more distinct than its neighbor, and the open end of its U-shape points toward the west. These dust clouds stand out by blocking the light from thousands of background stars in a rich section of the Milky Way.

This page and the one opposite pay tribute to Edward Emerson Barnard (1857–1923), the American astronomer whose photographic skills

revolutionized the study of bright and dark nebulae in the late 19th and early 20th centuries. He did much of his ground-breaking work at Lick Observatory on Mount Hamilton outside San Jose, California.

In addition to Barnard's E, he discovered Barnard's Star in Ophiuchus (the star with the highest proper motion of any known), Barnard's Galaxy (NGC 6822) in Sagittarius, Barnard's Loop in Orion, the California Nebula (NGC 1499) in Perseus, a number of comets, and the last moon in the solar system found visually, Jupiter's Amalthea.

Barnard's E
Also known as:
Barnard 142–3
Constellation: Aquila the Eagle
Right ascension: 19h40m
Declination: 10°31'
Magnitude: —
Apparent size: 60'x40'
Distance: 2,000 light-years

Barnard's Galaxy and the Little Gem Nebula

Although Barnard's Galaxy (NGC 6822) in Sagittarius glows faintly against the background sky, it does show up through small scopes under excellent conditions. LOCAL GROUP GALAXIES SURVEY TEAM/NOIRLAB/NSF/AURA

The Little Gem Nebula (NGC 6818) appears brighter than Barnard's Galaxy because its 9th-magnitude light concentrates in a much smaller area. ADAM BLOCK

While scanning the skies in northeastern Sagittarius in 1884 with a 5-inch refractor, master observer E.E. Barnard found—something. It turned out to be a small irregular galaxy located within our Local Group of galaxies some 1.5 million light-years from Earth.

Barnard's Galaxy (NGC 6822) can be tricky to find because it doesn't lie near any bright star. Perhaps your best starting point is magnitude 3.1 Beta (β) Capricorni in neighboring Capricornus. From there, slide 9° west and pick up an asterism of three 5th-magnitude stars (including 54 and 55 Sagittarii) that forms a 1°-long arc. From the northernmost star, slew your telescope 0.7° north-northeast and search for a misty patch about one-half the size of the Full Moon and only slightly brighter than the background star field. It will take averted vision to spot it in a small telescope.

If you train a 6-inch or larger telescope on the galaxy, you also might want to hunt for the Little Gem Nebula (NGC 6818), a 9th-magnitude planetary nebula located just 0.7° north-northwest of Barnard's Galaxy. It spans just 22", so it has a much higher surface brightness than the galaxy, which makes it a better target through bigger scopes. Still, I have observed both objects with an 80mm refractor as well as larger instruments. It's a challenge, but it can be done under excellent conditions with the right eyepiece—otherwise it wouldn't be in this book!

The Little Gem
Also known as: NGC 6818
Constellation: Sagittarius the Archer
Right ascension: 19h44m
Declination: –14°09'
Magnitude: 9.4
Apparent size: 22"x15"
Distance: 6,000 light-years

Barnard's Galaxy
Also known as: NGC 6822, Caldwell 57
Constellation: Sagittarius the Archer
Right ascension: 19h45m
Declination: –14°48'
Magnitude: 9.3
Apparent size: 16'x14'
Distance: 1.5 million light-years

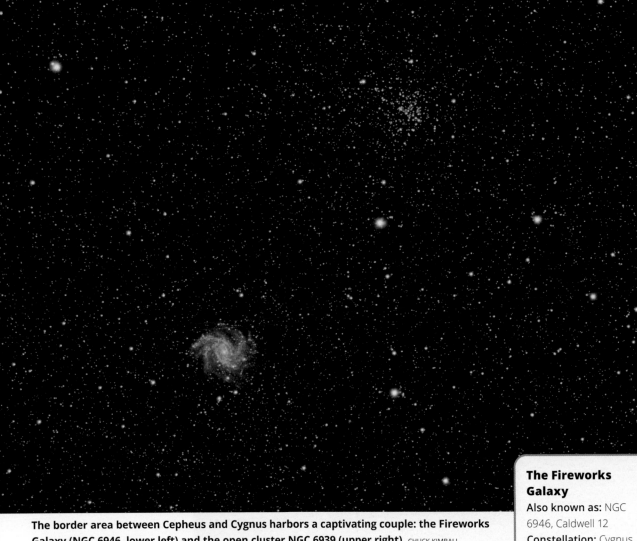

The border area between Cepheus and Cygnus harbors a captivating couple: the Fireworks Galaxy (NGC 6946, lower left) and the open cluster NGC 6939 (upper right). CHUCK KIMBALL

The Fireworks Galaxy and NGC 6939

An interesting vista awaits observers on the border between Cygnus and Cepheus. A chance alignment places the distant Fireworks Galaxy (NGC 6946) next to one of the Milky Way's own star clusters, NGC 6939.

To find this dynamic duo, first locate magnitude 2.5 Alpha (α) Cephei and magnitude 3.4 Eta (η) Cephei 3.9° to its west-southwest. Then, use your binoculars and continue in a southwesterly direction for about half the distance between those stars.

Although both deep-sky objects shine brightly enough to see through binoculars, they show up better with a telescope and low-power eyepiece. An 80mm refractor under a dark sky easily reveals the cosmic pair.

NGC 6939 is an open star cluster that lies 5,900 light-years from Earth, well within the confines of our galaxy. A small scope resolves some cluster members against a background haze formed by many unresolved stars. Spiral galaxy NGC 6946 lies 0.6°

southeast of the cluster. Through a small telescope, it appears as a round, nearly featureless glow owing to its distance of about 22 million light-years. NGC 6946 spans about one-third the size of the Full Moon and sports a small, bright core. The galaxy gets its common name from the fact that 10 supernovae have occurred in this galaxy since the beginning of the 20th century, more than in any other galaxy. This compares with the Milky Way's average rate of one or two each century.

The Fireworks Galaxy
Also known as: NGC 6946, Caldwell 12
Constellation: Cygnus the Swan
Right ascension: 20h35m
Declination: 60°09'
Magnitude: 8.9
Apparent size: 11'
Distance: 22 million light-years

NGC 6939
Also known as: Melotte 231
Constellation: Cepheus the King
Right ascension: 20h32m
Declination: 60°39'
Magnitude: 7.8
Apparent size: 7'
Distance: 5,900 light-years

The Veil Nebula

Let's head to southern Cygnus the Swan and track down the remnants of a star that exploded about 10,000 years ago. The softly glowing gas of the Veil Nebula forms a supernova remnant—all that's left of a giant star that began life with about 20 times the Sun's mass. Observers once considered the Veil a difficult object to see visually in small amateur telescopes, but modern optics make this a frequently visited summer object. Still, I consider this a challenge object because it can be tough to spot under less-than-ideal conditions without an OIII or a UHC filter.

The nebula consists of several different strands. The western section is easier to track down. Start with 1st-magnitude Deneb (Alpha [α] Cygni)—the Summer Triangle's northernmost star and the one that marks the Swan's tail—then drop 11° south to magnitude 2.5 Epsilon (ε) Cygni. From Epsilon, head 3.3° due south to 4th-magnitude 52 Cygni. If you focus your telescope on this star, the Veil's western segment appears as a wispy ribbon seemingly entangled with the star. (Note that 52 Cygni is a foreground object unrelated to the nebula.) The north end of this 1°-long ribbon tapers to a sharp point.

The Veil's brighter eastern section lies some 2.5° northeast of 52 Cygni. The wide-field 80mm scope I used for many of the descriptions in this book delivers great views of the nebula. Coupled with a 30mm wide-angle eyepiece (an 82° apparent field of view), this scope provides a 4.9° field of view; with a 22mm eyepiece (a 68° apparent field of view), the field measures 3.0°. Either choice lets you see the entire 3° circular extent of the Veil. Use higher powers or a larger scope to tease out detail in the nebula's filamentary arcs.

Although the Veil Nebula doesn't rank among the brightest deep-sky objects, I think you will find it one of the more memorable and mesmerizing targets to explore in the summer sky.

The Veil Nebula
Also known as: the Cygnus Loop, the Witch's Broom, Caldwell 33 (NGC 6992/5), Caldwell 34 (NGC 6960)
Constellation: Cygnus the Swan
Right ascension: 20h56m
Declination: 31°43'
Magnitude: 7.0
Apparent size: 210'x160'
Distance: 2,100 light-years

The Veil Nebula supernova remnant in Cygnus spans 3° because the gas expelled by the stellar explosion has spread 65 light-years in all directions during the past 10,000 years. STEVE CANNISTRA

The North America
and Pelican nebulae

The distinctive shape of the North America Nebula (NGC 7000, left) stands out in the Cygnus Milky Way; the less impressive Pelican Nebula (IC 5067/70, right) lies just 1° to its west. JASON WARE

With the spine of the Milky Way running through Cygnus the Swan, it shouldn't surprise you to find plenty of deep-sky objects within its borders. Many of these are the glowing clouds where gas and dust condense to form new stars. One of the finest lurks near the top of the Northern Cross asterism (the tail of the Swan): the North America Nebula (NGC 7000). One look with the right telescope and filters will make it clear how this stellar nursery got its common name.

Locating NGC 7000 is a breeze. Start at Deneb, the brightest star in Cygnus. Through binoculars or a small telescope with a low-power eyepiece, center this 1st-magnitude star and then move 2° directly east. You should see a bright region of the Milky Way along with obvious dark dust lanes. Check the object's orientation in your field—if you are facing south and viewing with binoculars, the North America Nebula will be properly oriented. The dark dust clouds that form the continent's Gulf of Mexico stand out best.

Binoculars provide the top method for "discovering" NGC 7000. I recommend using an 8x42, 7x50, 10x50, or 11x80 model (a tripod helps steady the view). But your best view will come from a small telescope with an OIII filter and an eyepiece that provides a 3° or larger field. The wide-field 80mm telescope with a 22mm or 35mm 2-inch eyepiece will deliver an exceptional vista showing the entire object's familiar shape with enough contrast to see it directly.

The Pelican Nebula (IC 5067/70) stands 1° west of NGC 7000. Although the nebula shows up nicely in photos, it appears much fainter than the North America and shows little detail. The odds are against seeing the shape of a bird in its contours.

The North America should enjoy a status equivalent to the best nebulae in the Milky Way. Although the Lagoon (M8) and Orion (M42) nebulae shine brighter, neither climbs as high as NGC 7000 for northern observers, and thus we must view them through thick layers of air.

The North America Nebula

Also known as: NGC 7000, Caldwell 20
Constellation: Cygnus the Swan
Right ascension: 20h59m
Declination: 44°20'
Magnitude: 6.0
Apparent size: 120'x100'
Distance: 2,600 light-years

The Pelican Nebula

Also known as: IC 5067/70
Constellation: Cygnus the Swan
Right ascension: 20h48m
Declination: 44°22'
Magnitude: 8.0
Apparent size: 60'x50'
Distance: 1,800 light-years

Gamma Cygni region and Albireo

Cygnus the Swan holds many wonderful deep-sky objects. While the Veil and the North America nebulae are spectacular, the area around the constellation's center star—magnitude 2.2 Gamma (γ) Cygni—holds its own when viewed from a dark site.

Starting with the naked eye, you'll notice the Milky Way appears brighter in Cygnus. That's because you are looking along the length of a spiral arm 90° from the galaxy's center. Also notice a split in the Milky Way that runs roughly through the middle of Cygnus down to Sagittarius. This is the Great Rift—the visible manifestation of all the dust clouds lying in our galaxy's plane.

Take your binoculars and scan the area around Gamma. You should notice two small 7th-magnitude open star clusters: NGC 6910 just 0.5° to the star's north-northeast and M29 1.8° south-southeast of the star. The faint reflection nebula NGC 6914 lies 2° north of Gamma and the equally faint emission nebula IC 1318 resides mostly east and north of the star.

If you use a wide-field telescope with an OIII filter, the most striking feature of this region is the sheer number of dark nebulae—dust clouds fill the whole field. It's kind of like looking at a loose bouquet of white carnations with dark amaranth leaves filling the spaces in between the flowers.

If you really get serious about tracking down large emission nebulae, you may want to consider getting a pair of matching OIII filters that can thread over the eyepieces of your binoculars (after you unthread the rubber eyecups). The wide field that binoculars provide often give more impressive views than the

Albireo (Beta Cygni) forms one of the sky's most colorful double stars. The brighter component glows distinctly yellow while the fainter one has a bluish cast. ALAN DYER

The small open star cluster M29 in central Cygnus glows at magnitude 6.6 and lies just 2° south of the bright star Gamma Cygni. RICHARD HAMMAR

Rich pockets of bright and dark nebulae surround magnitude 2.2 Gamma Cygni, the second-brightest star in Cygnus the Swan. ADAM BLOCK

narrower fields most telescopes offer.

While you're in Cygnus, drop down to the constellation's southwestern corner and the bright star Albireo (Beta [β] Cygni) that marks both the head of the Swan and the base of the Northern Cross. Find it in your telescope at low power but then switch to a medium power, about 40x

to 50x. Albireo resolves into two stars—a magnificent example of a double star. If you look even closer, you'll find the two stars have different colors. The brighter one shines yellow while the fainter one sports a bluish hue (or greenish depending on your eyesight). The color contrast arises because the blue star is some 14,400° F (8,000° C) hotter than its companion. About half the sky's stars belong to double or multiple-star systems.

Albireo
Also known as: Beta (β) Cygni
Constellation: Cygnus the Swan
Right ascension: 19h31m
Declination: 27°58'
Magnitude: 3.1 and 5.1
Apparent size: 34" separation
Distance: 420 light-years

Gamma Cygni region
Also known as: NGC 6622
Constellation: Cygnus the Swan
Right ascension: 20h19m
Declination: 40°44'
Magnitude: 10.0
Apparent size: 50'
Distance: 1,800 light-years (star)

M22
Also known as:
NGC 6656
Constellation:
Sagittarius the Archer
Right ascension:
18h36m
Declination: −23°54'
Magnitude: 5.1
Apparent size: 32'
Distance: 10,000 light-years

Messier globulars in Sagittarius

Let's finish our tour of the summer sky with a spin around Sagittarius the Archer to visit all seven of the globular star clusters Charles Messier cataloged in the constellation. Use a good star chart to get your bearings and you should be set. We'll start in northern Sagittarius and work our way clockwise around the constellation.

M22 (NGC 6656): The brightest globular in Sagittarius shines at magnitude 5.1 and should be easy to see with the unaided eye on a dark night. Like all the Messier globulars in Sagittarius, M22 shows up easily with direct vision through an 80mm wide-field telescope. This cluster appears as large as the Full Moon. It looks big and bright in part because it lies only 10,000 light-years from Earth. You should be able to resolve some of its stars at higher power.

M28 (NGC 6626): This globular lies 2.9° west-southwest of M22. It appears about one-third the size and noticeably dimmer than its neighbor. Why the difference? At a distance of 18,000 light-years, it resides nearly twice as far away.

M69 (NGC 6637): Located inside the Archer's Teapot asterism, this globular lies 29,000 light-years away and glows at magnitude 7.6. Although easy to spot through binoculars or a finder scope, it appears less grainy than the previous clusters. Its great distance means an 80mm scope can't resolve stars.

M70 (NGC 6681): Scan 2.5° due east of M69 and you'll land on M70. Like its neighbor, this magnitude 7.9 globular shines across 29,000 light-years. Of historical interest, this was the object astronomers Alan Hale and Thomas Bopp were viewing when they stumbled across Comet C/1995 O1 (Hale-Bopp), which would become a great comet in 1997.

M54 (NGC 6715): Glowing at magnitude 7.6, this globular lies a whopping 86,000 light-years away. To appear as big and bright as it does, it must be a giant. Astronomers have discovered that it actually resides beyond the boundaries of our galaxy, in the Sagittarius Dwarf Galaxy.

M55 (NGC 6809): This nice globular shines at magnitude 6.3 across a distance of 18,000 light-years. This ranks as the second-best object on our list.

M75 (NGC 6864): Located on Sagittarius' eastern border, this globular resides 68,000 light-years from Earth. You might expect it to be smaller and fainter than most of the other objects on this list—and you'd be right. It glows at magnitude 8.5 and appears only 20 percent the size of M22.

Milky Way globulars orbit the center of our galaxy in a giant halo and rarely come close to the galactic core, though M70 (NGC 6681) is an exception to this rule. ESA/HUBBLE & NASA

M28 (NGC 6626) glows at magnitude 6.8 and appears about one-third the size of M22. HAREL BOREN

M54 (NGC 6715) resides 86,000 light-years from Earth in the Sagittarius Dwarf Galaxy. ESO

Magnitude 7.6 M69 (NGC 6637) lies far enough from Earth that its stars prove difficult to resolve through small scopes. STARHOPPER/CC BY-SA-4.0

Sagittarius abounds with bright globulars, but only M22 outshines magnitude 6.3 M55 (NGC 6809). DANIEL VERSCHATSE

M75 (NGC 6864) stands out as perhaps the least impressive of Messier's globulars in Sagittarius, but it's still worth viewing through small telescopes. ADAM BLOCK

AUTUMN SKY

■ Autumn marks a season of transition, and the night sky responds with an impressive diversity of astronomical objects. Fall brings shorter days and longer, cooler nights offering bright galaxies, brilliant star clusters, dazzling nebulae, and vivid shells of dying stars. Even the Milky Way is changing from its summer dominance toward the more subtle yet still intricate winter Milky Way.

The autumn sky section of this book covers objects in the night sky with right ascensions between 21 and 3 hours. This quadrant of the sky features the constellation Pegasus the Winged Horse and its prominent asterism, the Great Square of Pegasus, above the horizon every night at 10 P.M. local daylight time from roughly September through November.

The fall sky holds the brightest galaxies, the largest planetary nebula, and some of the finest star clusters you can see from Earth, a fitting prelude to the magnificent star clusters and emission nebulae of the winter sky.

Autumn nights can get quite nippy, so you'll want to break out some of your winter observing gear—but at least you shouldn't need to worry about observing in a snowdrift. Your chances of great observing weather are pretty good at this time of year as well. In fact, October has some of the highest percentages of clear night skies across the Northern Hemisphere.

The fall sky is notable for its variety of objects. **My autumn Top Ten list** (again in order of total brightness) reflects this range: **1** The Andromeda Galaxy (M31); **2** The Double Cluster (NGC 869 and NGC 884), two adjoining naked-eye open star clusters in Perseus; **3** The Pinwheel Galaxy (M33) in Triangulum; **4** The Iris Nebula (NGC 7023) in Cepheus, a standout reflection nebula; **5** M15 (NGC 7078) in Pegasus, the season's finest globular cluster; **6** M52, a glorious open cluster with a nearby emission nebula in Cassiopeia; **7** The Helix Nebula (NGC 7293), a huge, low-surface-brightness planetary nebula in Aquarius; **8** The Silver Dollar Galaxy (NGC 253) in Sculptor; **9** The Pacman Nebula (NGC 281) in Cassiopeia; and **10** The Skull Nebula (NGC 246) in Cetus.

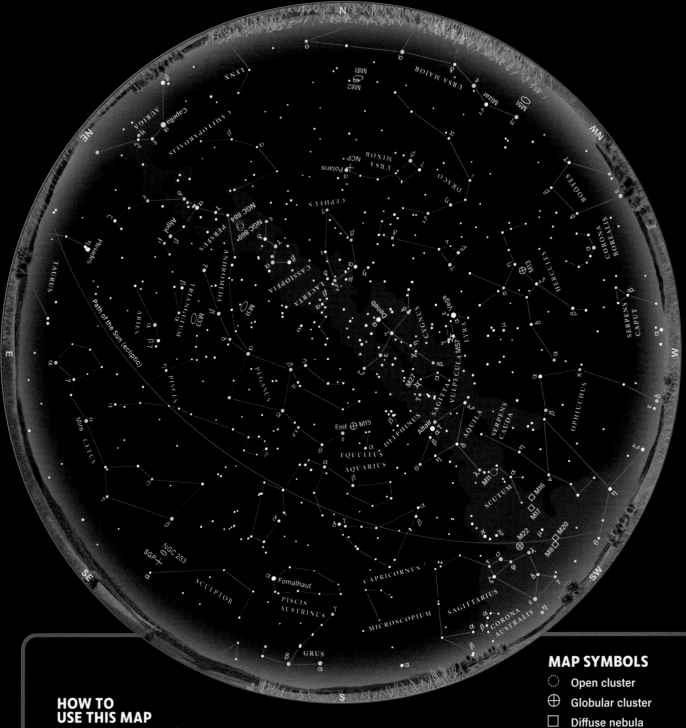

HOW TO USE THIS MAP

This map portrays the sky as seen near 35° north latitude. Located inside the border are the cardinal directions and their intermediate points. To find stars, hold the map overhead and orient it so one of the labels matches the direction you're facing. The stars above the map's horizon now match what's in the sky.

The all-sky map shows how the sky looks at:

1 A.M. September 15
11 P.M. October 15
8 P.M. November 15

STAR COLORS

A star's color depends on its surface temperature.

- The hottest stars shine blue
- Slightly cooler stars appear white
- Intermediate stars (like the Sun) glow yellow
- Lower-temperature stars appear orange
- The coolest stars glow red
- Fainter stars can't excite our eyes' color receptors, so they appear white unless you use optical aid to gather more light

MAP SYMBOLS

- ⬭ Open cluster
- ⊕ Globular cluster
- ☐ Diffuse nebula
- ✧ Planetary nebula
- ⬯ Galaxy

STAR MAGNITUDES

- ● Sirius
- ● 0.0
- ● 1.0
- ● 2.0
- · 3.0
- · 4.0
- · 5.0

AUTUMN SKY OBJECTS

PAGE	OBJECT NAME	CONSTELLATION	TYPE	R.A.	DEC.	MAG.	SIZE
126	The Saturn Nebula (NGC 7009)	Aquarius	Planetary nebula	21h04m	–11°22'	8.0	1.6'x0.4'
127	The Iris Nebula (NGC 7023)	Cepheus	Reflection nebula	21h02m	68°10'	6.8	18'
128	M15 (NGC 7078)	Pegasus	Globular cluster	21h30m	12°10'	6.2	18'
129	M2 (NGC 7089)	Aquarius	Globular cluster	21h33m	–0°49'	6.5	16'
130-131	The Elephant Trunk Nebula (IC 1396)	Cepheus	Emission nebula	21h39m	57°30'	3.5	170'
132	The Spare Tyre Nebula (IC 5148)	Grus	Planetary nebula	22h00m	–39°23'	11	2'
133	NGC 7209	Lacerta	Open cluster	22h05m	46°29'	7.7	25'
133	NGC 7243	Lacerta	Open cluster	22h15m	49°54'	6.4	21'
134-135	The Helix Nebula (NGC 7293)	Aquarius	Planetary nebula	22h30m	–20°50'	7.3	20'
136-137	The Bubble Nebula (NGC 7635)	Cassiopeia	Emission nebula	23h21m	61°12'	10	15'
136-137	M52 (NGC 7654)	Cassiopeia	Open cluster	23h25m	61°36'	6.9	13'
138	The Blue Snowball Nebula (NGC 7662)	Andromeda	Planetary nebula	23h26m	42°32'	8.3	2.2'
139	NGC 7822	Cepheus	Emission nebula	0h01m	67°25'	5.7	100'
140	The Whale Galaxy (NGC 55)	Sculptor	Spiral galaxy	0h15m	–39°12'	7.9	32'x6'
141	NGC 147	Cassiopeia	Dwarf spheroidal galaxy	0h33m	48°31'	9.5	13'x8'
141	NGC 185	Cassiopeia	Dwarf spheroidal galaxy	0h39m	48°20'	9.2	12'x10'
142-143	The Andromeda Galaxy (M31)	Andromeda	Spiral galaxy	0h43m	41°16'	3.4	180'x65'
144	The Skull Nebula (NGC 246)	Cetus	Planetary nebula	0h47m	–11°52'	8	3.8'
144	NGC 255	Cetus	Spiral galaxy	0h48m	–11°28'	11.7	3'x2'
145	The Claw Galaxy (NGC 247)	Cetus	Spiral galaxy	0h47m	–20°46'	9.2	21'x7'
146	The Silver Dollar Galaxy (NGC 253)	Sculptor	Spiral galaxy	0h48m	–25°17'	7.6	28'x7'
146	NGC 288	Sculptor	Globular cluster	0h53m	–26°35'	8.1	13.8'
147	The Pacman Nebula (NGC 281)	Cassiopeia	Emission nebula	0h53m	56°37'	7.8	30'x35'
148	M103 (NGC 581)	Cassiopeia	Open cluster	1h33m	60°39'	7.4	6'
149	NGC 654	Cassiopeia	Open cluster	1h44m	61°53'	6.5	5'

PAGE	OBJECT NAME	CONSTELLATION	TYPE	R.A.	DEC.	MAG.	SIZE
149	NGC 659	Cassiopeia	Open cluster	1h44m	60°40'	7.9	6'
149	NGC 663	Cassiopeia	Open cluster	1h46m	61°13'	7.1	16'
150-151	The Pinwheel Galaxy (M33)	Triangulum	Spiral galaxy	1h34m	30°40'	5.7	73'x45'
152	The Phantom Galaxy (M74)	Pisces	Spiral galaxy	1h37m	15°47'	9.4	10'
153	NGC 752	Andromeda	Open cluster	1h57m	37°48'	5.7	75'
154	The Double Cluster (NGC 869 & NGC 884)	Perseus	Open clusters	2h21m	57°08'	4.3/4.4	30' each
155	The Squid Galaxy (M77)	Cetus	Spiral galaxy	2h43m	–0°01'	8.9	7'x6'
156-157	The Heart Nebula (IC 1805)	Cassiopeia	Emission nebula	2h33m	61°28'	6.5	60'
156-157	The Soul Nebula (IC 1848)	Cassiopeia	Emission nebula	2h51m	60°25'	6.5	60'x30'
158	NGC 1097	Fornax	Spiral galaxy	2h46m	–30°16'	9.2	9.3'x6.3'

The Saturn Nebula

For our first fall object, we're going to look at a small classic—small in the sense of appearing tiny in the sky, not for a lack of brightness or visual beauty. The Saturn Nebula (NGC 7009) is a planetary nebula, a dying star in the process of expelling its outer layers. By chance, this expanding gas cloud resembles a slightly out-of-focus Saturn.

To find NGC 7009, first track down Aquarius the Water-bearer. This large and dim constellation lies south of Pegasus and east of Aquila, or north and east of Capricornus. Western Aquarius features the constellation's two brightest stars, Alpha (α) and Beta (β) Aquarii, which both shine at magnitude 2.9. Some 8° southwest of Beta lies magnitude 4.5 Nu (ν) Aquarii. The Saturn Nebula resides 1.3° due west of Nu.

NGC 7009 appears small, with its oval disk measuring about 25" across. Either scan the area with enough power to see the disk, probably at least 50x, or look for a slightly greenish star and then magnify it. (Of course, finding small planetaries becomes a lot easier if you have a go-to mount or some other sort of computer-aided or robotic telescope.) The nebula glows brightly enough to see with binoculars, but it will look starlike.

With good seeing and 100 power or more in an 80mm telescope, you will get a hint of the "rings," though it takes a 6-inch or larger scope to really see them. Those rings extend the nebula 15" from each end. With good seeing, you will be able to view NGC 7009's complex structure, with a brighter core and at least one fainter, outer shell.

A white dwarf lies at the nebula's center. This star radiates ultraviolet light that powers the nebula's emission. In theory this magnitude 11.5 star should be visible with an 80mm scope, but it's a challenge.

The Saturn Nebula
Also known as:
NGC 7009, Caldwell 55
Constellation: Aquarius the Water-bearer
Right ascension: 21h04m
Declination: –11°22'
Magnitude: 8.0
Apparent size: 1.6'x0.4'
Distance: 4,000 light-years

The Saturn Nebula (NGC 7009) lies in southwestern Aquarius just 1.3° west of Nu Aquarii. You can find it at low power, but you'll need higher powers to see the "rings." DANIEL VERSCHATSE

The Iris Nebula

The Iris Nebula (NGC 7023) dwells in Cepheus the King. It's a bright reflection nebula that's part of an open star cluster and an easy target with an 80mm scope. To find this object, start at magnitude 3.2 Beta (β) Cephei. With binoculars or a telescope at low power, move 3.3° southwest. The star cluster associated with the Iris Nebula shines at magnitude 6.8 and spans about half the size of the Full Moon.

In a telescope, you'll first see a 7th-magnitude star. A glowing white area surrounds this star. You are looking at a reflection nebula—where dust reflects the star's light—with a delicate and irregular shape that resembles an eye's iris. You can see it with direct vision

Cepheus the King offers observers one of the sky's best and brightest examples of a reflection nebulae: the Iris Nebula (NGC 7023). T.A. RECTOR/UNIVERSITY OF ALASKA ANCHORAGE, H. SCHWEIKER/WIYN AND NOIRLAB/NSF/AURA

but, because it's a reflection nebula, a nebular filter will not improve the view.

The Iris ranks among the brightest reflection nebulae in the sky. It lies about 1,400 light-years from Earth.

The Iris Nebula
Also known as:
NGC 7023, Caldwell 4
Constellation: Cepheus the King
Right ascension: 21h02m
Declination: 68°10'
Magnitude: 6.8
Apparent size: 18'
Distance: 1,400 light-years

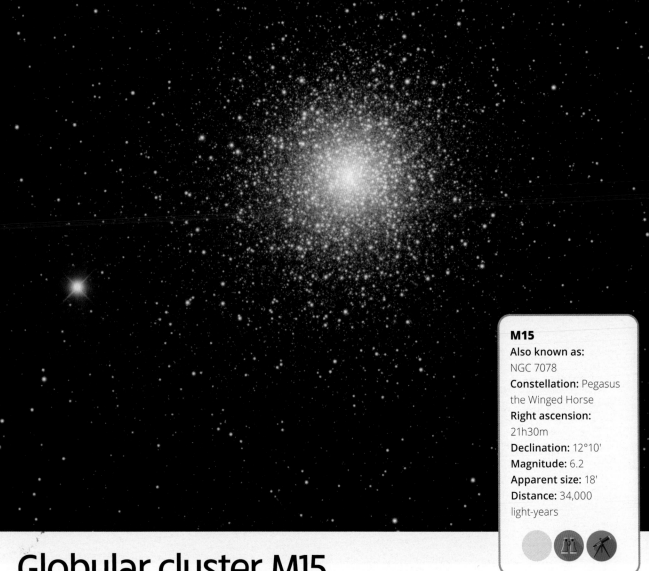

Globular cluster M15

Drop south from Cepheus past Cygnus and you'll find the constellation Pegasus the Winged Horse and its conspicuous asterism, the Great Square. From the square's southwestern corner, magnitude 2.5 Alpha (α) Pegasi, hop southwest to magnitude 3.5 Theta (θ) Pegasi and then northwest to magnitude 2.4 Enif (Epsilon [ε] Pegasi). Continue 4.2° in the same northwesterly direction past Epsilon and you will land on M15 (NGC 7078). This globular cluster appears bright and large enough to show up easily as a fuzzy ball of stars in binoculars.

Through an 80mm wide-field telescope at high power, you should start to resolve a few stars at the edges of the magnitude 6.2 cluster (though true resolution takes a 6-inch scope). M15 has a more condensed core than most globulars.

The globular star cluster M15 (NGC 7078) calls Pegasus home. Astronomers estimate it to be at least 12 billion years old, making it one of the oldest globulars known. T.A. RECTOR (UNIVERSITY OF ALASKA ANCHORAGE) AND H. SCHWEIKER (WIYN AND NOIRLAB/NSF/AURA)

This shows up as a more jewel-like appearance if you compare it with a classic like M13 in Hercules or M5 in Serpens. Astronomers say the cluster has undergone core collapse, and it may host a black hole weighing some 4,000 solar masses in its core. M15 also harbors the planetary nebula Pease 1, but it appears stellar even in large scopes. This was the first planetary discovered in a globular. Another planetary, NGC 7094, lies 1.8° north-northeast of M15, but you'll need an 8-inch or larger instrument to catch it.

Globular cluster M2

If you enjoy observing globular clusters, summer is your time of year. But autumn is no desert—not only is M15 in Pegasus a standout, but so is M2 in Aquarius the Water-bearer. You have two convenient methods for tracking down this globular. Option 1 is to head 4.8° due north of magnitude 2.9 Beta (β) Aquarii. Option 2 applies if you recently viewed M15. M2 stands about 13° south of Pegasus' bright globular.

Magnitude 6.5 M2 glows a bit fainter than M15, but it's still an easy catch through binoculars or a small telescope at low power. It lies about 37,000 light-years from Earth, a few thousand light-years farther than M15, and appears 10 percent smaller than its northern neighbor. You won't resolve M2 with an 80mm scope, but it should appear mottled or grainy at higher powers.

Aquarius holds the wonderful globular cluster M2 (NGC 7089). Look for it in binoculars nearly 5° due north of 3rd-magnitude Beta Aquarii. ESA/HUBBLE & NASA, G. PIOTTO ET AL.

M2
Also known as:
NGC 7089
Constellation: Aquarius the Water-bearer
Right ascension:
21h33m
Declination: –0°49'
Magnitude: 6.5
Apparent size: 16'
Distance: 37,000 light-years

The Elephant Trunk Nebula

Head back to Cepheus the King and look for the shape of a house like a child might draw, with four stars in a square representing the main structure and a fifth one to the north marking the tip of the roof. Magnitude 2.5 Alpha (α) Cephei and magnitude 3.3 Zeta (ζ) Cephei form the house's base. Just south of a line between these two stars lies Mu (μ) Cephei. Commonly called Herschel's Garnet Star, this star varies from magnitude 3.4 to 5.1 over a 730-day period and glows with an intensely reddish hue.

Now look at Mu with binoculars when Mu stands near the meridian. You should see a faint glowing area about 3° wide mostly south of the star. This is IC 1396—a vast star-forming region some 3,000 light-years away near the edge of the Milky Way's glow—and home of the Elephant Trunk Nebula (IC 1396A). (Some observers report seeing it with the naked eye and averted vision, but that's a truly impressive achievement.)

In a wide-field 80mm telescope with an OIII filter, a 2-inch diagonal, and a 35mm, 68° eyepiece delivering a massive 4.9° field of view, you can clearly see the outline of the nearly circular emission patch with direct vision when the transparency is good.

It only took me about 15 years to get there. I first moved to a remote ranch with a dark sky in 2003. I tried to observe IC 1396 with brute force, even using telescopes with apertures as high as 30 inches, but had no luck seeing the big picture. Finally, in frustration, I measured the fields of view I was getting and realized that I was using too large a telescope with too much power to see the entire object. That's when I dug out the 80mm wide-field refractor and got amazing views.

The Elephant Trunk Nebula

Also known as: IC 1396
Constellation: Cepheus the King
Right ascension: 21h39m
Declination: 57°30'
Magnitude: 3.5
Apparent size: 170'
Distance: 3,000 light-years

IC 1396, the bright nebula surrounding the dark Elephant Trunk Nebula, shows up with the naked eye at the southern edge of Cepheus where the King borders Cygnus. IC 1396 is also visible in binoculars, but you need a really wide-angle telescope to see the whole thing. JOEL SHORT

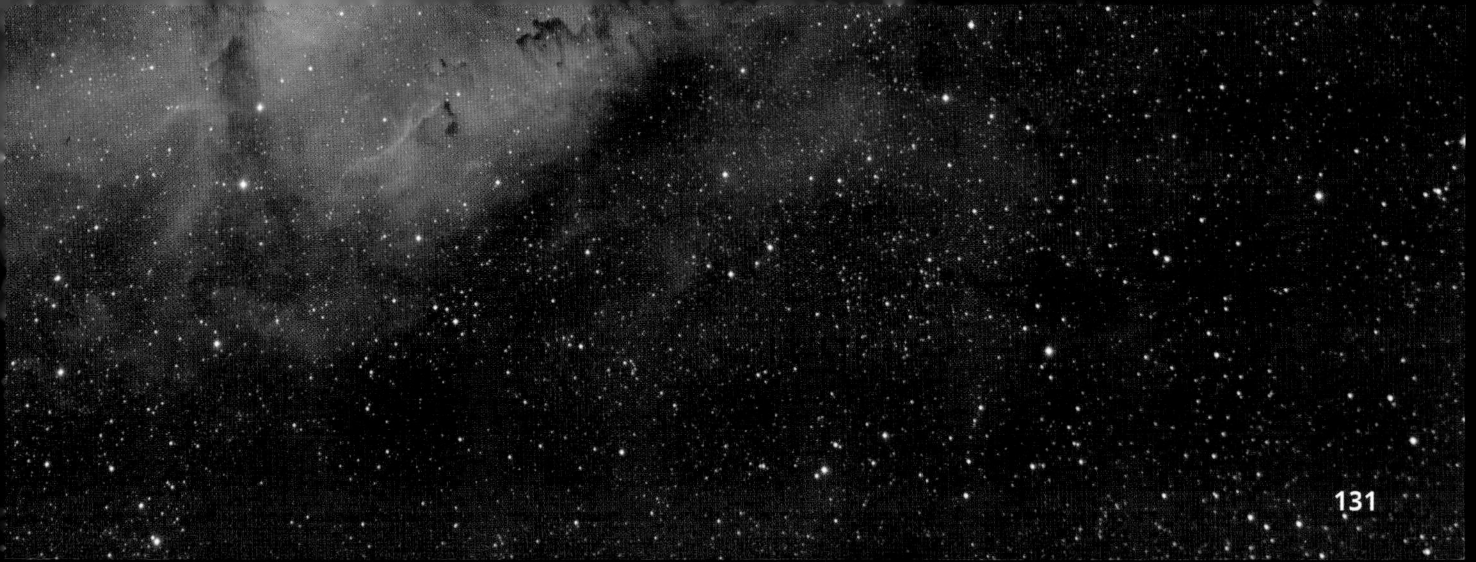

The Spare Tyre Nebula

Also known as: IC 5148
Constellation: Grus the Crane
Right ascension: 22h00m
Declination: –39°23'
Magnitude: 11
Apparent size: 2'
Distance: 3,800 light-years

Let's take a plunge into the southern constellation Grus the Crane. The big bird lies south of Piscis Austrinus and its 1st-magnitude luminary, Fomalhaut, which itself stands south of Aquarius. Zero in on magnitude 4.5 Lambda (λ) Gruis and push your telescope 1.3° west to find the Spare Tyre Nebula (IC 5148).

This object proves a real challenge for an 80mm telescope, and I would suggest using a 6-inch or larger scope. The object is one of the sky's larger planetary nebulae, spanning 2', and an astonishing object to come across.

IC 5148 has an annular appearance similar to the Ring Nebula, though slightly larger. Its common name comes from its shape, while the uncommon spelling derives from British English.

Open clusters NGC 7209 and NGC 7243

Now get ready to explore a place few people have gone, Lacerta the Lizard. This constellation's brightest star, Alpha (α) Lacertae, glows dimly at magnitude 3.8. You can find Lacerta sandwiched between Cygnus to the west and Pegasus to the south.

Despite its uninspiring appearance, Lacerta possesses two noteworthy deep-sky objects, the open star clusters NGC 7209 and NGC 7243, hiding among the stars of the Milky Way. Once you identify the constellation, find Alpha and then the slightly fainter stars 4 and 2 Lacertae. Using binoculars, head 2° west from either Alpha or 4 Lacertae and you should see NGC 7243 shining at magnitude 6.4. To find magnitude 7.7 NGC 7209, scan 2.7° west of 2 Lacertae.

The two clusters are easy to see with binoculars or an 80mm telescope despite the rich Milky Way background. My wide-field 80mm refractor at low power fully resolves both objects. NGC 7243 appears about two-thirds the size of the Full Moon and looks somewhat elongated. NGC 7209 looks a little larger than its neighbor but is more loosely packed.

NGC 7209
Constellation: Lacerta the Lizard
Right ascension: 22h05m
Declination: 46°29'
Magnitude: 7.7
Apparent size: 25'
Distance: 4,100 light-years

NGC 7243
Also known as: Caldwell 16
Constellation: Lacerta the Lizard
Right ascension: 22h15m
Declination: 49°54'
Magnitude: 6.4
Apparent size: 21'
Distance: 2,800 light-years

The seldom-visited northern constellation Lacerta holds two nice open star clusters, NGC 7209 and NGC 7243, visible through binoculars. NGC 7209 (top) shines slightly dimmer than its companion. ANTHONY AYIOMAMITIS

The Helix Nebula

One of fall's best sights lies in the southern part of the constellation Aquarius the Water-bearer. The easiest way to find the Helix Nebula (NGC 7293) is to start at Aquarius' Water Jar asterism just south of the constellation's border with Pegasus. Then scan 20° south of the asterism's eastern end to find magnitude 5.2 Upsilon (υ) Aquarii. The Helix lies 1.2° west of Upsilon. Or, you can locate Upsilon and the Helix 11° northwest of 1st-magnitude Fomalhaut.

The Helix is a beast of a planetary. It spans two-thirds the diameter of a Full Moon, making it one of largest planetary nebulae. It appears big because it's about 10,000 years old and thus has had a lot of time to expand, and because it lies only 650 light-years from Earth.

NGC 7293 shows up easily in 8x42 or larger binoculars if the transparency is good and the Moon is not out. Because the nebula has a fairly low surface brightness, however, it's a little hard to locate if the transparency isn't top notch.

You will see a mostly uniform disk through an 80mm telescope, but its annularity appears noticeable on clear nights. As with most of the non-stellar objects in this book, the larger the telescope you use,

the better the view, at least until the field of view gets narrow enough in a large telescope that you can see only a part of the object at one time. Because the Helix is an emission nebula, either an OIII or UHC filter will boost the contrast and make it far more conspicuous.

I had the opportunity to view the Helix from a high-altitude observing site in the California Sierras on an exceptional night and got an astonishing view of some of the cometary knots in the nebula's interior. These thick balls of material occur when fast-moving gas in the expanding envelope catches up with slower-moving gas and create tail-like structures that point away from the central star.

> ### The Helix Nebula
> **Also known as:**
> NGC 7293, Caldwell 63
> **Constellation:** Aquarius the Water-bearer
> **Right ascension:** 22h30m
> **Declination:** –20°50'
> **Magnitude:** 7.3
> **Apparent size:** 20'
> **Distance:** 650 light-years
>
>

The wonderful Helix Nebula (NGC 7293) ranks among the finest planetaries in the sky. This big and fairly faint object shows up in binoculars and small telescopes; an OIII filter will help you see it. NASA/NOAO/ESA/THE HUBBLE HELIX NEBULA TEAM/M. MEIXNER (STSCI)/T.A. RECTOR (NRAO)

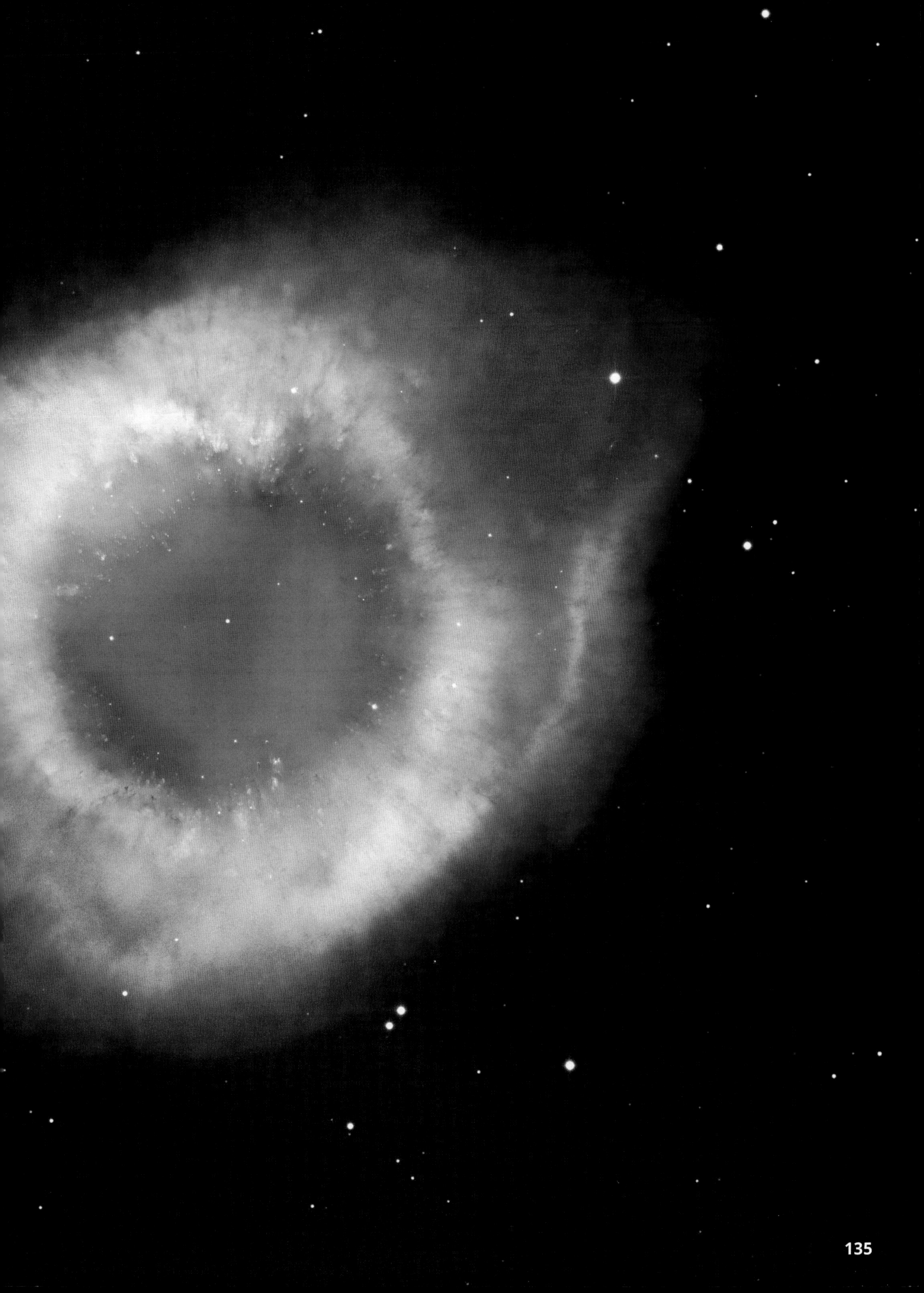

The Bubble Nebula and M52

You might suppose that a constellation located in the Milky Way's plane, like Cassiopeia the Queen, would hold lots of open star clusters and bright nebulae. And you'd be right. The Queen holds nearly 10 percent of the open clusters known in our galaxy, so you'd expect its best would be a stunner. M52 (NGC 7654) delivers nicely on this promise. To find it, first locate Alpha (α) and Beta (β) Cassiopeiae, the two 2nd-magnitude stars that form the western end of the W-shaped constellation. Draw an imaginary line from Alpha to Beta, extend it westward an equivalent distance, and the cluster will pop into view through binoculars.

M52 glows at magnitude 6.9, so it falls just below naked-eye visibility for most people. The cluster shines through an 80mm wide-field scope, even at low power. It appears bright and highly condensed, yet still resolves nicely.

A second worthy object stands 0.6° southwest of M52. The Bubble Nebula (NGC 7635) rates as a top astro-imaging target even though it's difficult to see visually. Use low power on a moonless night when Cassiopeia hangs near the meridian and the sky's transparency is excellent. Even with an OIII filter, NGC 7635 appears exceedingly faint and emerges sporadically as a partial outline of a circle with averted vision. The Bubble's northern edge glows brighter than the rest and

constitutes a sighting of this object. The nebula's spherical shape arises from a fierce stellar wind blowing from a superhot star positioned north of the object's center. (The star appears off-center because the stellar wind encounters more material to the north, slowing its progress.)

NGC 7635 demonstrates how much Earth's atmosphere impacts the visibility of faint objects. If you can see the nebula when it lies near the meridian, come back to it a few hours later when it's about 45° high and try again. The increased atmospheric absorption often will render the object invisible.

Western Cassiopeia holds a delightful open star cluster, M52 (NGC 7654), and an elusive emission region holding the Bubble Nebula (NGC 7635) 0.6° to the southwest. Try to catch the Bubble using an OIII filter when it is high in a clear sky. HUNTER WILSON

Open cluster M52 (NGC 7654) shines at 7th magnitude and makes a tempting target through both binoculars and telescopes of all sizes. NOIRLAB/NSF/AURA

The Bubble Nebula (NGC 7635) explodes with detail when imaged with the 4-meter telescope on Kitt Peak. T.A. RECTOR/UNIVERSITY OF ALASKA ANCHORAGE, H. SCHWEIKER/WIYN AND NOIRLAB/NSF/AURA

The Bubble Nebula

Also known as:
NGC 7635, Caldwell 11
Constellation:
Cassiopeia the Queen
Right ascension:
23h21m
Declination: 61°12'
Magnitude: 10
Apparent size: 15'
Distance: 7,900
light-years

M52

Also known as:
NGC 7654
Constellation:
Cassiopeia the Queen
Right ascension:
23h25m
Declination: 61°36'
Magnitude: 6.9
Apparent size: 13'
Distance: 4,400
light-years

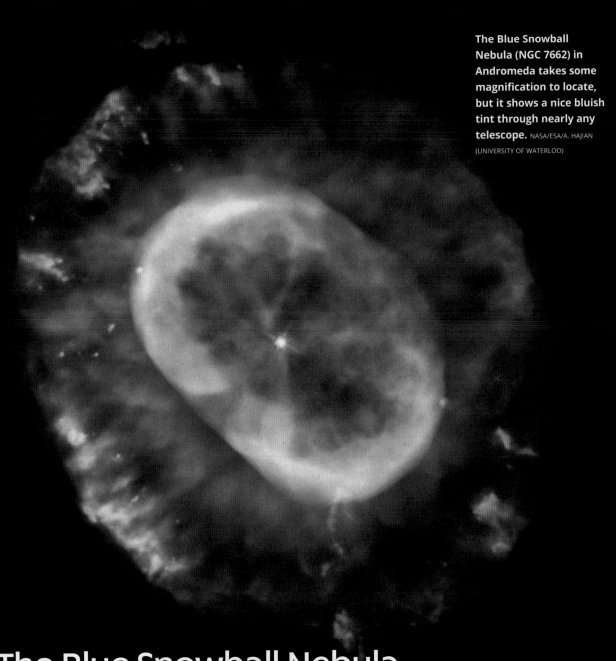

The Blue Snowball Nebula

Andromeda the Princess lies north and east of Pegasus. Two 4th-magnitude stars, Kappa (κ) and Iota (ι) Andromedae, lie near the constellation's western edge. Scan 2.4° west-southwest of Iota and you'll notice a bluish "star." Magnify this object 50 to 100 times and you'll see the small, round, bluish ball of a planetary nebula.

The Blue Snowball Nebula (NGC 7662) proves a little tough to find with a wide-angle telescope. By definition, such scopes deliver low-power views, so you're attempting to identify a small object without magnifying it much. You're probably better off searching with slightly more power than usual.

The view through an 80mm wide-field scope shows a blue ball at about 50x. When searching at 29x, the planetary shows just a tiny disk barely different from a star. Since most planetaries appear small, I didn't include many in this book. If you want to explore more of these captivating objects, buy a larger telescope. Your views of small planetaries and galaxies will benefit immensely. If you make the plunge and purchase a big scope, keep your small wide-angle instrument for wide-field targets and easy transport.

The Blue Snowball Nebula
Also known as:
NGC 7662, Caldwell 22
Constellation:
Andromeda the Princess
Right ascension:
23h26m
Declination: 42°32'
Magnitude: 8.2
Apparent size: 2.2'
Distance: 2,500 light-years

Emission nebula NGC 7822

Let's travel north again, to Cepheus the King and the region where the constellation's border with neighboring Cassiopeia forms a stairstep pattern. Our target here is the star-forming region NGC 7822. To find it, start at either magnitude 3.5 Iota (ι) Cephei and scan 7.5° east or at magnitude 2.3 Beta (β) Cassiopeiae and travel 9.5° north. Use your telescope and a low-power eyepiece to sweep it up.

In my wide-field 80mm scope and a 22mm eyepiece (which yields 23x), I can clearly see a ribbon of nebulosity trending east-west across the field of view. An OIII filter helps, but NGC 7822 shows up without the filter. The nebula is churning out hot stars at a great rate. The hottest of these newborn suns radiate lots of ultraviolet light that excites the surrounding hydrogen gas to glow red.

Another large patch of nebulosity, Cederblad 214, sits just 1.5° south of NGC 7822. The more southerly object does not show up in my 80mm scope, and some star charts do not even map it. When I attempted to locate it, the conditions were not highly favorable, so it will be a challenge for a future observing session with a bigger scope.

NGC 7822
Constellation: Cepheus the King
Right ascension: 0h01m
Declination: 67°25'
Magnitude: 5.7
Apparent size: 100'
Distance: 2,900 light-years

The emission nebula NGC 7822 appears as a horizontal ribbon at the top of this image. To its south lies a more circular emission patch cataloged as Cederblad 214. GEORGE GREANEY

The Whale Galaxy

Sometimes referred to as the Whale Galaxy, NGC 55 is a nearly edge-on spiral or irregular galaxy in the southern constellation Sculptor. ESO

South of Cetus the Whale lies the constellation Sculptor the Sculptor, which conveniently harbors the deep-sky object known as the Whale Galaxy (NGC 55). The easiest way to locate this gem is to start at 1st-magnitude Fomalhaut in neighboring Piscis Austrinus. From there, sweep 11° southeast and pick up magnitude 4.4 Beta (β) Sculptoris. NGC 55 lies 8° east and 1° south of Beta.

The Whale Galaxy is an edge-on barred spiral—though some scientists classify it as an irregular—that lies nearly 7 million light-years from Earth. Although it shines quite brightly, at 8th magnitude, its brilliance is muted for Northern Hemisphere observers because it lies well south and so we must view it through thick layers of atmosphere.

NGC 55 has an uneven thickness that gave rise to its nickname. Through an 80mm refractor, it appears as a fairly uniform, long streak that may show some mottling, but you'll need a larger scope to see much detail. Images reveal that NGC 55's eastern flank is more dynamic than its quieter western side. The galaxy forms a pair with NGC 300 some 8° to the east. The two likely interact gravitationally, and they appear to reside in a void between our Local Group of galaxies and the Sculptor Group.

NGC 55 is but one of two cetaceans roaming the celestial sea. We covered the northern sky's version, NGC 4631 in Canes Venatici, on page 73.

The Whale Galaxy
Also known as: NGC 55, Caldwell 72
Constellation: Sculptor the Sculptor
Right ascension: 0h15m
Declination: –39°12'
Magnitude: 7.9
Apparent size: 32'x6'
Distance: 6.9 million light-years

A pair of Local Group galaxies

Let's jump northward to Cassiopeia the Queen and track down a pair of galaxies. First locate magnitude 2.3 Alpha (α) Cassiopeiae, then drop 8° south to magnitude 4.5 Omicron (o) Cassiopeiae. Using a telescope and low-power eyepiece, look 1° to 2° west for the soft glows of NGC 185 (Caldwell 18) and NGC 147 (Caldwell 17). The two appear side by side and just 1° apart.

A telescope provides excellent views. They are easy targets for an 80mm telescope at low power, appearing as slightly oval patches with no detail.

NGC 147 and NGC 185 are dwarf galaxies belonging to our Local Group and orbit the much larger Andromeda Galaxy (M31).

NGC 147

Also known as:
Caldwell 17
Constellation:
Cassiopeia the Queen
Right ascension: 0h33m
Declination: 48°31'
Magnitude: 9.5
Apparent size: 13'x8'
Distance: 2.3 million light-years

NGC 185

Also known as:
Caldwell 18
Constellation:
Cassiopeia the Queen
Right ascension: 0h39m
Declination: 48°20'
Magnitude: 9.2
Apparent size: 12'x10'
Distance: 2.2 million light-years

NGC 147 (top) and NGC 185 are both dwarf spheroidal galaxies and appear close enough to each other to view through a wide-field scope at the same time. Both are satellites of the much larger Andromeda Galaxy (M31). NASA/ESA/A. FERGUSON (UNIVERSITY OF EDINBURGH, INSTITUTE FOR ASTRONOMY)

The Andromeda Galaxy

The Andromeda Galaxy (M31) and its satellites form one of the greatest vistas in the entire sky. Binoculars and low-power telescopes show this deep-sky treasure best.

To find the galaxy, first locate the constellation Andromeda the Princess. Start with the star at the northeastern corner of the Great Square of Pegasus. Despite belonging to this asterism, the magnitude 2.1 star belongs to Andromeda and is named Alpha (α) Andromedae. From Alpha, head northeast along an arc of 4th-magnitude stars for about two binocular fields. There you will find magnitude 3.9 Mu (μ) Andromedae. If you now hop 4° northwest from Mu, you'll see a fuzzy patch with your naked eye. Aim your

The Andromeda Galaxy (M31) is a naked-eye object from a dark-sky site when the Moon is not visible. You can appreciate its size—3° long, or six Full Moons— even in small binoculars.

BILL SCHOENING/VANESSA HARVEY/REU PROGRAM/NOIRLAB/NSF/AURA

This close-up shows NGC 205—a satellite galaxy of M31. If you look carefully, even a small telescope will reveal two satellite galaxies nearly superimposed on M31's disk.

KPNO/NOIRLAB/NSF/AURA/ADAM BLOCK

The Andromeda Galaxy

Also known as: M31, NGC 224

Constellation: Andromeda the Princess

Right ascension: 0h43m

Declination: 41°16'

Magnitude: 3.4

Apparent size: 180'x65'

Distance: 2.5 million light-years

binoculars at this area and you'll spy a large elliptical glow: the Andromeda Galaxy.

You should have little trouble spotting M31 with your unaided eye from a dark site. (In fact, at 2.5 million light-years away, it is the most distant object most people can see with the naked eye.) And any binoculars will give you a good view. If you then point your wide-field, low-power telescope at that fuzzy spot, you will start to see details. First, look for its bright and smooth central region with spiral arms wrapping around it. Then, examine the arms for irregularities. You also should see a small elliptical ball off to the galaxy's northwest side. This is NGC 205, a dwarf galaxy that orbits M31. Higher magnifications should

reveal a second satellite galaxy, the dwarf M32 (NGC 221) on the opposite side of M31's core. M32 is harder to make out because it lies closer to M31's core and appears superimposed on the larger galaxy's disk.

In a small telescope like my wide-field 80mm, M31's core and disk appear fairly uniform, but as you look at it with larger scopes, you'll start to see dust lanes and other features. You can visually spy star clouds, star clusters, globulars, and emission regions—M31 becomes an intricate star field like the Milky Way.

If you relax and soak in the view for a while, you will start to see further details, just as you would

for any object you study for long. Can you see the yellow glow of the core and the bluish tint in the arms? These colors reflect the stellar populations in these regions—the core holds mostly old and cool stars while the spiral arms harbor hot young stars.

Do you like what you see? It will only get better. The Andromeda Galaxy is approaching the Milky Way at around 250,000 mph. You can expect a rather big bump in the night when our galaxies collide in about 4 billion to 5 billion years. No, you won't see any change in your life, but big things are in store for our distant descendants.

The Skull Nebula (NGC 246) is a large planetary nebula in Cetus. If you look carefully at the bright "star" to its upper left, it turns out to be the small round galaxy NGC 255. ESO/DSS2

The Skull Nebula and NGC 255

The fall sky features many constellations related to water. The largest of these, Cetus the Whale, swims south below the stars of Pisces the Fish. Our next two targets lie in the southwestern corner of Cetus, in the great sea creature's tail. Locate Cetus' brightest star, magnitude 2.0 Beta (β) Ceti, and then look about 7° (one binocular field) north for a chain of three 5th-magnitude stars: Phi^1 (φ^1), Phi^2 (φ^2), and Phi^3 (φ^3) Ceti.

The Skull Nebula (NGC 246) resides 1.4° south-southwest of Phi^2. Although bright enough to see with binoculars, it looks like an ordinary star. Turn an 80mm telescope on the Skull and it shows up easily with direct vision, though you'll likely need to use more power to identify it as non-stellar. Such a set-up shows a small and mostly featureless disk. The markings within the expanding shell of this dying star that hint at the nebula's common name become apparent with a 12-inch scope.

If you use a 6-inch or larger instrument, look 0.4° north-northeast of the Skull for barred spiral galaxy NGC 255. This small galaxy appears nearly round and glows at magnitude 11.7, so it lies beyond the range of an 80mm scope.

The Skull Nebula
Also known as:
NGC 246, Caldwell 56
Constellation: Cetus the Whale
Right ascension: 0h47m
Declination: –11°52'
Magnitude: 8
Apparent size: 3.8'
Distance: 1,800 light-years

NGC 255
Constellation: Cetus the Whale
Right ascension: 0h48m
Declination: –11°28'
Magnitude: 11.7
Apparent size: 3'x2'
Distance: 71 million light-years

The Claw Galaxy

Second-magnitude

Beta (β) Ceti turns out to be the best jumping-off point for another spiral galaxy in Cetus the Whale. Using binoculars, scan 2.9° south-southeast of this star and you should see a faint patch of light elongated north-south. The Claw Galaxy (NGC 247) belongs to the Sculptor Group of galaxies and lies 12 million light-years from Earth.

When you turn a small telescope on this nebulosity, your first reaction may be to wonder why the magnitude 9.2 object looks so faint. Blame its low surface brightness. The galaxy's luminosity spreads out over a large area, reducing the light that comes from any individual sector. This makes it hard to pick out from the background sky, though you will see an elongated shape characteristic of edge-on spirals.

Larger telescopes show the Claw has a complex, disturbed interior. One mysterious feature is an apparent void in the northern part of the galaxy's disk. Although stars do exist in this gap, they are older and dimmer than the bright blue stars and pinkish emission nebulae elsewhere in the disk.

> **The Claw Galaxy**
> **Also known as:**
> NGC 247, Caldwell 82
> **Constellation:** Cetus the Whale
> **Right ascension:** 0h47m
> **Declination:** –20°46'
> **Magnitude:** 9.2
> **Apparent size:** 21'x7'
> **Distance:** 12 million light-years

The Claw Galaxy (NGC 247) is a nearly edge-on spiral in Cetus. (East appears up in this image.) This is one of the closest major spiral galaxies to Earth. ESO

The Silver Dollar Galaxy (NGC 253) is a jewel of the southern sky. The galaxy appears elongated through binoculars; a telescope may reveal mottling. ESO/INAF-VST

Look 1.8° southeast of NGC 253 and you'll see the glow of globular cluster NGC 288. It appears unresolved in small scopes. ESA/HUBBLE & NASA

The Silver Dollar Galaxy
Also known as:
NGC 253, Caldwell 65
Constellation: Sculptor the Sculptor
Right ascension: 0h48m
Declination: –25°17'
Magnitude: 7.6
Apparent size: 28'x7'
Distance: 12 million light-years

NGC 288
Constellation: Sculptor the Sculptor
Right ascension: 0h53m
Declination: –26°35'
Magnitude: 8.1
Apparent size: 13.8'
Distance: 29,000 light-years

The Silver Dollar Galaxy and globular NGC 288

Starting at the Claw Galaxy (NGC 247) I covered on the previous page, scan with binoculars 4.5° due south. Along the way, you will have passed across the border separating Cetus from Sculptor the Sculptor and alighted on another elongated nebula: the Silver Dollar Galaxy (NGC 253).

This galaxy, sometimes referred to simply as the Sculptor Galaxy, is more than four times brighter than NGC 247 and spans a length nearly equal to the Full Moon's diameter. It is the prime member of the Sculptor Group of galaxies, the closest such group to our own Local Group at a distance of some 12 million light-years. Unfortunately, it lies at a declination of –25°, keeping it relatively low in the sky for northern observers. On the plus side,

it lies only 2° from our galaxy's south pole, so we see it with no intervening dust.

At low power in an 80mm refractor, the galaxy appears as a fairly bright, elongated oval. You may see some mottling in the disk and a concentration of light at the center. And you should see why the galaxy got its common name—it looks like an edge-on silver dollar. A low-power view

also should reveal a small fuzzy ball 1.8° southeast of the galaxy. This is the unresolved Milky Way globular cluster NGC 288.

Larger scopes reveal some of NGC 253's structure, including dust lanes, two major spiral arms, and a nuclear bulge. The Sculptor Galaxy spans about 90,000 light-years and is creating new stars at a furious rate.

The Pacman Nebula

After spending the previous few pages on objects south of the celestial equator, let's move back north to the constellation Cassiopeia the Queen. To find our next target, first locate magnitude 2.3 Alpha (α) Cassiopeiae, the southernmost star in the W-shaped asterism that dominates the constellation. If you sweep the area about 2° east of the star, you'll see evidence of a star cluster and the faint light of an emission nebula. Point a small scope to this spot and at low power you should see a Moon-sized glow. (If you don't, an OIII filter will help.) This is the Pacman Nebula (NGC 281), a region of hydrogen gas energized by several hot stars packed into the embedded cluster IC 1590.

The Pacman rates as an easy target through an 80mm telescope from a dark-site. Even in a scope of this size you'll see an asymmetry in the nebula—it seems to be missing a chunk out of its southwestern side. It gives the object a resemblance to the popular 1980s video game character, hence its common name. In reality, a dark nebula creates the missing piece by obscuring part of the emission. The Pacman Nebula adds interest to an already fascinating constellation. It surprises me that it did not make the Caldwell Catalog—though I realize Patrick Moore must have considered an astronomical number of possibilities.

The nebula glows at about magnitude 7.8 from a distance of 10,000 light-years. It lies in the Perseus Arm of the Milky Way.

The Pacman Nebula (NGC 281) sits just south of Cassiopeia's W shape and east of the bright star Alpha Cassiopeiae. This circular patch of emission nebulosity is a star-forming region; try looking for it with binoculars when it lies high in the sky. T.A. RECTOR/UNIVERSITY OF ALASKA ANCHORAGE AND WIYN/AURA/NSF

The Pacman Nebula
Also known as:
NGC 281
Constellation:
Cassiopeia the Queen
Right ascension: 0h53m
Declination: 56°37'
Magnitude: 7.8
Apparent size: 30'x35'
Distance: 10,000 light-years

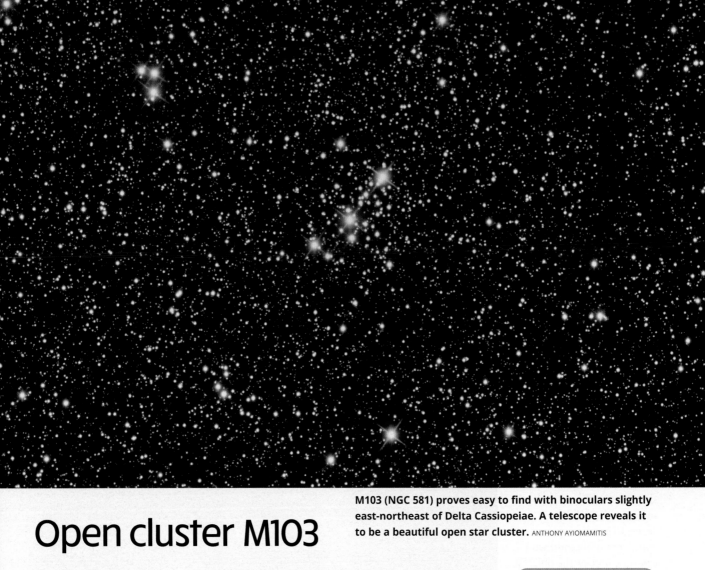

Open cluster M103

M103 (NGC 581) proves easy to find with binoculars slightly east-northeast of Delta Cassiopeiae. A telescope reveals it to be a beautiful open star cluster. ANTHONY AYIOMAMITIS

Cassiopeia the Queen is a playground for those with small telescopes. Bright star clusters and faint emission nebulae fill the constellation with wonders that can keep you busy throughout a long autumn night. If you point your binoculars at magnitude 2.7 Delta (δ) Cassiopeiae, the second star in from the eastern side of the

constellation's W shape, you also will see a faint triangular glow 1° east-northeast of Delta. Now point your low-power telescope at this target and you'll find a nice open star cluster with a mix of bright and faint stars that reminds me of the shape of a Christmas Tree.

Your binoculars may not be able to resolve this

cluster, which renders it looking like a gaseous nebula, but this is only the combined light of many faint stars packed in an area only 6' across. Cassiopeia is a rich hunting ground for star clusters. If you have a wide-angle telescope, three more open clusters lie within 2° of M103. I'll tackle those on the next page.

M103
Also known as:
NGC 581
Constellation:
Cassiopeia the Queen
Right ascension: 1h33m
Declination: 60°39'
Magnitude: 7.4
Apparent size: 6'
Distance: 10,000 light-years

A queen's trio of open clusters

As I mentioned on the previous page, three more open star clusters lie in M103's vicinity. This trio comprises NGC 654, NGC 659, and NGC 663 (Caldwell 10). If you haven't just explored M103, the easiest way to find this threesome is to start at magnitude 2.7 Delta (δ) Cassiopeiae on the eastern side of Cassiopeia the Queen's W-shaped asterism. Using binoculars, look between 2° and 3° east and northeast of Delta (or 1° to 2° beyond M103) and you should spot the north-south chain of clusters.

Starting at the north end, you'll find magnitude 6.5 NGC 654, magnitude 7.1 NGC 663, and finally magnitude 7.9 NGC 659. All should be apparent through a wide-field 80mm telescope, with NGC 659 the faintest but still nice. As you might expect from its Caldwell designation, NGC 663 stands as the best of the group.

All these clusters lie roughly 6,000 light-years from Earth and were born between 15 and 40 million years old. They appear to be members of the Cassiopeia OB8 stellar association and reside in the Perseus Arm of the Milky Way.

In eastern Cassiopeia, just east of M103, lies a chain of three smaller open star clusters. NGC 654 (top) and NGC 663 (bottom) appear in this image; NGC 659 lies farther south. All are easy targets for binoculars and small telescopes. BERNHARD HUBL

NGC 654
Constellation:
Cassiopeia the Queen
Right ascension: 1h44m
Declination: 61°53'
Magnitude: 6.5
Apparent size: 5'
Distance: 5,900 light-years

NGC 659
Constellation:
Cassiopeia the Queen
Right ascension: 1h44m
Declination: 60°40'
Magnitude: 7.9
Apparent size: 6'
Distance: 6,300 light-years

NGC 663
Also known as:
Caldwell 10
Constellation:
Cassiopeia the Queen
Right ascension: 1h46m
Declination: 61°13'
Magnitude: 7.1
Apparent size: 16'
Distance: 6,400 light-years

The Pinwheel Galaxy

Our next target resides in the inconspicuous constellation Triangulum the Triangle. To find the Pinwheel Galaxy (M33), start just north in Andromeda. The Princess' main pattern consists of two curving lines of stars. The middle star in the southern arc is magnitude 2.1 Beta (β) Andromedae. M33 lies 7° southeast of this star. Or simply look an equal distance to the southeast of Beta as the Andromeda Galaxy (M31) is to the star's northwest.

If you're outside on the darkest of nights, you might be able to make out the faint glow of M33 without optical aid. Binoculars show a sizable patch of nebulosity that spans more than twice the Full Moon's size.

With a small wide-angle scope, the Pinwheel shows as a faint oval glow. M33 has one of the lowest surface brightnesses of any Messier object, so it won't knock your socks off. Compare it in your mind— or by switching back and forth—with M31. While the Andromeda Galaxy has a large and bright core, the Pinwheel has a small nucleus. Increasing the magnification won't help much. In fact, it might hurt because the image will dim.

Opposite Beta Andromedae from M31 is the fabulous Pinwheel Galaxy (M33). Although the spiral arms won't be apparent through binoculars or small scopes, larger amateur instruments show them. T.A. RECTOR (NRAO/AUI/NSF AND NOIRLAB/NSF/AURA) AND M. HANNA (NOIRLAB/NSF/AURA)

The Pinwheel Galaxy

Also known as: The Triangulum Galaxy, M33, NGC 598
Constellation: Triangulum the Triangle
Right ascension: 1h43n
Declination: 30°40'
Magnitude: 5.7
Apparent size: 73'x45'
Distance: 3.0 million light-years

In 6-inch and larger telescopes, you should see a brighter patch in M33's disk. This is NGC 604, one of the biggest emission nebulae in the nearby universe. NASA/ESA/M. DURBIN, J. DALCANTON, AND B.F. WILLIAMS (UNIVERSITY OF WASHINGTON)

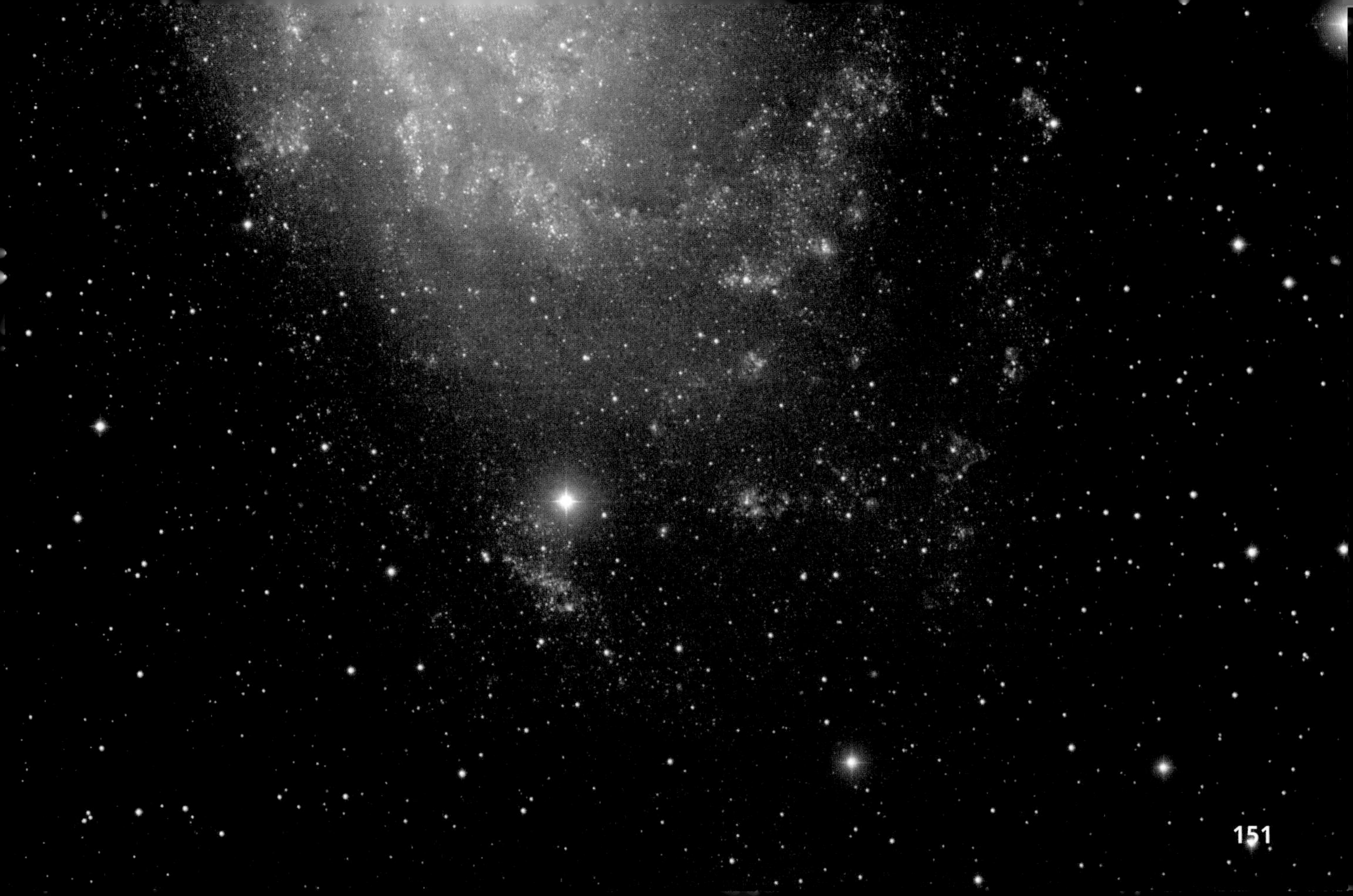

The Phantom Galaxy

Spiral galaxy M74 lies in eastern Pisces near its border with Aries. Nicknamed the Phantom Galaxy for its low surface brightness, M74 proves to be the most difficult Messier object to observe. R. JAY GABANY

Immediately south of Andromeda and Pegasus lies Pisces the Fish, another of fall's watery constellations. Although Pisces holds a number of NGC objects, it has only one that made the Messier Catalog. The Phantom Galaxy (M74) is a grand design spiral, meaning it has prominent and well-defined spiral arms. On a clear night during the dark of the Moon, trace out the pattern of the Fish and hone in on the magnitude 3.6 star Eta (η) Piscium. Using binoculars or a telescope with a low-power eyepiece, look 1.3° east-northeast to spot the Phantom.

In an 80mm wide-angle telescope, you'll want to observe M74 on a dark transparent night when it passes near the meridian. M74's common name came about because the galaxy has an exceptionally low surface brightness that causes many people to miss it. M74 has the lowest surface brightness of any Messier object except M101. M101 is about three times larger, however, so it appears more noticeable. M74's low surface brightness comes in part from its nearly face-on orientation—its disk tilts 85° to our line of sight. An 80mm telescope shows M74 as a modest-sized, circular milky spot with a small central condensation. I consider the Phantom to be a challenge object.

M74 is a loosely wound spiral, a larger clone of M33 that looks smaller because it lies about 10 times farther away. By coincidence, about 10 galaxy groups and clusters reside within 10° or so of M74. The Phantom is the brightest member of the M74 Group, which includes up to six fainter galaxies.

The Phantom Galaxy

Also known as: M74, NGC 628

Constellation: Pisces the Fish

Right ascension: 1h37m

Declination: 15°47'

Magnitude: 9.4

Apparent size: 10'

Distance: 30 million light-years

Open cluster NGC 752

Our next target lies in Andromeda the Princess nearly due north of M74, but at a distance of 22°, there are easier ways to find it. First locate the two curving lines of stars that form Andromeda's major pattern, then zero in on the easternmost star in the southern arc: magnitude 2.2 Gamma (γ) Andromedae. Using binoculars, scan about 4° south and slightly west and you easily should pick out open star cluster NGC 752 (Caldwell 28). This magnitude 5.7 cluster spans more than twice the size of the Full Moon. If you have good eyesight and an excellent dark-sky location, you should be able to spot NGC 752 without optical aid.

In my wide-field 80mm telescope, Caldwell 28 appears loosely packed with 50 to 70 stars spread across its large diameter. All told, astronomers have cataloged more than 250 cluster members. These are mostly old stars—at least for an open cluster—with ages of more than 1 billion years. I also like to think of NGC 752 as an "unexpected" cluster, one that appears in a rather empty portion of space and proves delightful and pleasing.

While you're in this area, linger a bit longer on 2nd-magnitude Gamma Andromedae. This is perhaps the finest double star in the autumn sky. A small telescope easily splits the two main members in two—one with a distinct yellow tint and the other sporting a blue-green hue—separated by about 10". Although star colors usually appear subtle, put two contrasting suns together in a single eyepiece view and they become more vivid. Interestingly, the fainter star actually consists of three stars, making Gamma a four-star system.

NGC 752
Also known as:
Caldwell 28
Constellation:
Andromeda the Princess
Right ascension: 1h57m
Declination: 37°48'
Magnitude: 5.7
Apparent size: 75'
Distance: 1,500 light-years

The open star cluster NGC 752 lies some 4° south of the wonderful double star Gamma Andromedae. Binoculars reveal a loose, large cluster more than twice the Full Moon's size. ANTHONY AYIOMAMITIS

The Double Cluster

The far northwestern sector of Perseus the Hero holds one of autumn's finest vistas. To locate the Double Cluster (NGC 869 and NGC 884), scan with your binoculars between the "W" of Cassiopeia and the head of Perseus. If you want more precise directions, draw a line from magnitude 2.2 Gamma (γ) Cassiopeiae to magnitude 2.7 Delta (δ) Cassiopeiae in the W and then extend the line by twice the distance between the stars.

When you gaze in this direction with your naked eye on a dark night, you should see the cluster pair as a brighter fuzzy area in the Milky Way. This is a fine binocular target and

especially nice to view with a wide-field telescope.

In a wide-field 80mm scope, you will notice two bright open star clusters standing side by side. Both NGC 869 (to the west) and NGC 884 (on the east) appear well-resolved despite the rich background of stars. Just 0.5° separates the centers of the two. NGC 869 boasts a higher concentration of stars, with nearly three dozen 10th-magnitude and brighter suns packed inside. NGC 884 spreads out a bit more but contains more bright stars.

This is one target where larger telescopes and higher magnifications don't really hurt the view. Pumping up

The Double Cluster (NGC 869 and NGC 884) shows up to the naked eye as a brighter patch of haze in the Milky Way in northern Perseus. Binoculars and small telescopes reveal two rich, bright open star clusters standing side-by-side against a rich background. N.A. SHARP/NOIRLAB/NSF/AURA

the power won't let you see both clusters simultaneously, but the remaining cluster explodes into more stars. A 4-inch scope also reveals plenty of color among the stars, with healthy mixtures of blue, yellow, and red gems vying for attention.

The two clusters lie about 7,500 light-years from Earth in the Perseus Arm of the Milky Way Galaxy. NGC 884 possesses a mass of about 3,700 Suns while NGC 869 contains 4,700 solar masses. Both clusters are youngsters on a galactic scale, only about 14 million years old. They

> **The Double Cluster**
> **Also known as:** h and χ Persei, NGC 869 and NGC 884, Caldwell 14
> **Constellation:** Perseus the Hero
> **Right ascension:** 2h21m
> **Declination:** 57°08'
> **Magnitude:** 4.3 and 4.4
> **Apparent size:** 30' each
> **Distance:** about 7,500 light-years
>
>

stand at the core of the Perseus OB1 stellar association of hot young stars.

The Squid Galaxy and companions

We spent some time several pages ago examining objects in the tail section of Cetus the Whale. Now let's turn our attention to the sea beast's head south of Aries and west of Taurus. Despite being the sky's fourth-largest constellation and having an abundance of fine deep-sky objects, Cetus contains only one Messier object—the Squid Galaxy (M77).

To find M77, start at magnitude 3.5 Gamma (γ) Ceti in the Whale's head and then drop 3.1° south-southwest to magnitude 4.1 Delta (δ) Ceti. Using binoculars or low power on a wide-field telescope, look 0.9° east-southeast of Delta. Binoculars may show only a smudge, but the telescope at higher power reveals a fairly bright oval elongated slightly in an east-west direction. The same set-up exposes a large nucleus and variations in the outer disk.

A prototypical Seyfert galaxy, M77 stands as one of the brightest (magnitude 8.9), closest (33 million light-years away), and best studied of its class. Such galaxies have exceptionally bright nuclei powered by supermassive black holes. (The one in M77 tips the scales at about 15 million solar masses.) Astronomers prize Seyferts because they are closer, lower-powered versions of quasars.

The Squid Galaxy (M77) in Cetus lies within 1° of Delta Ceti. A telescope reveals this spiral's bright nucleus, which characterizes Seyfert galaxies. NASA/ESA/A. VAN DER HOEVEN

The Squid Galaxy
Also known as: M77, NGC 1068
Constellation: Cetus the Whale
Right ascension: 2h43m
Declination: –0°01'
Magnitude: 8.9
Apparent size: 7'x6'
Distance: 33 million light-years

A large complex of emission nebulae and star clusters inhabit the southeastern corner of Cassiopeia. The brightest regions are the Heart Nebula (right) and the Soul Nebula (left). JACK NEWTON

The Heart and Soul nebulae

If you use binoculars or a wide-field telescope to observe the Double Cluster in Perseus (see page 154), swing north and slightly east some 5° into Cassiopeia the Queen and start your quest for the Heart and Soul nebulae (IC 1805 and IC 1848, respectively). First find magnitude 3.3 Epsilon (ε) Cassiopeiae and magnitude 3.8 Eta (η) Persei. Draw a line between these two stars, find the midpoint, and then scan about 1° northeast.

Look for the Heart and Soul nebulae on a moonless night with excellent transparency when they lie near the meridian. Also use a low-power eyepiece and preferably an OIII or UHC

filter. Although these objects appear bright, colorful, and detailed when imaged, they are a challenge to view visually. You probably will see them only as indistinct glows with averted vision in a scope.

This region divides nicely into two groups composed of multiple objects spread over nearly 5°. On the western edge of the western group, look for a bright knot of nebulosity designated NGC 896. It blends into IC 1805 and together they form the Heart Nebula. A bright star cluster at the center of IC 1805 energizes the hydrogen and causes it to glow. A 7th-magnitude star cluster, NGC 1027, shows up

through binoculars on the nebula's eastern edge.

A little more than 1° southeast of NGC 1027 lurks the eastern group, which the Soul Nebula dominates. The emission nebula appears large and diffuse, and it has a low surface brightness. Your best chance to see it comes with a wide-angle telescope with a low-power eyepiece and an OIII filter. On close inspection, the nebula breaks into two nearly separate regions that remind some observers of a baby's head and body. The embedded star cluster shows up more easily, particularly if you remove the filter. This entire region lies smack in the plane of the Milky Way.

The Heart Nebula
Also known as: IC 1805
Constellation:
Cassiopeia the Queen
Right ascension: 2h33m
Declination: 61°28'
Magnitude: 6.5
Apparent size: 60'
Distance: 5,500
light-years

The Soul Nebula
Also known as: IC 1848
Constellation:
Cassiopeia the Queen
Right ascension: 2h51m
Declination: 60°25'
Magnitude: 6.5
Apparent size: 60'x30'
Distance: 6,500
light-years

The Heart Nebula (IC 1805) combines glowing gas clouds and dark dust clouds, with a bright and energetic star cluster at its center. BOB AND JANICE FERA

The Soul Nebula (IC 1848) features both an emission region and a star cluster. Both it and the Heart Nebula are faint visually and show up best with a nebula filter. NEIL FLEMING

Spiral galaxy NGC 1097

South of Cetus lies the galaxy-rich constellation Fornax. Scan 2.2° north-northwest of Beta Fornacis to find the classic barred spiral galaxy NGC 1097. ESO

NGC 1097

Also known as:
Caldwell 67

Constellation: Fornax the Furnace

Right ascension: 2h46m

Declination: –30°16'

Magnitude: 9.2

Apparent size: 9.3'x6.3'

Distance: 48 million light-years

This book's final target is an isolated galaxy fairly deep in the southern sky among the stars of Fornax the Furnace. NGC 1097 (Caldwell 67) shines at magnitude 9.2 and shows up well in 10x50 binoculars under a dark sky.

To find Fornax, look southeast of Cetus. You should see the constellation's two brightest stars: magnitude 3.8 Alpha (α) and magnitude 4.5 Beta (β) Fornacis. Once you spot them, look for NGC 1097 2.2° to the north-northwest.

In an 80mm telescope, NGC 1097 appears elongated and should show a small, bright nuclear region. In much larger scopes, you'll start to see a Z-like pattern that results from a central bar and pronounced spiral arms.

NGC 1097 is another Seyfert galaxy, one with a supermassive black hole weighing some 140 million solar masses. The galaxy also has a fairly bright satellite just to its north that almost appears embedded in its outer portions. It glows at 13th magnitude, however, so you would need a 16-inch or bigger scope to see it. About two binocular fields southeast of NGC 1097 lies the Fornax Galaxy Cluster, which is well worth exploring with a larger scope. We covered one of its members, NGC 1365, in the spring sky section.

GLOSSARY (Deep-sky slang)

7x50 — Designation for a binocular with 7 power and a 50mm objective lenses.

Achromatic refractor — An improved design of Galileo's original, basic refractor telescope design that uses two lenses at its front; these objective lenses better correct for the color error introduced by simple lenses.

Aperture fever — An affliction commonly developed by deep-sky observers, the signature symptom being the overwhelming desire to see more in the nighttime sky by acquiring the largest telescope possible.

Apochromatic refractor — A refractor with lenses made with rare-earth glass that bend light without the color error of simple achromatic refractors. Often made with three or more lenses in the large objective lens at the front of the telescope. It performs much more like a reflector but is generally more expensive (twice or more) than a simple achromatic refractor.

Averted vision — The art of looking off-axis (slightly away from where you suspect an object) and be able to detect or see the object without staring directly at it. You eye is more sensitive to faint objects that are off-axis.

Bortle Scale — A 1 to 9 rating of the darkness of the sky; the lower the number, the better the sky quality. Class 1 are the darkest skies available on Earth; class 9 would be what you would find in the middle of a large city. See more about this scale by visiting the International Dark Sky Association (https://www.darksky.org/).

Caldwell Catalog — A catalog developed by English astronomer Sir Patrick Moore who attempted to catalog the next 109 objects Charles Messier would have or should have discovered. Generally, it's a good list of brighter deep-sky objects but less well-known than the Messier Catalog.

Cerro Tololo — A mountain in Chile where the U.S. National Science Foundation has a large observatory, the Cerro Tololo Interamerican Observatory. On the peak of the 7,241-foot-tall mountain is a 4-meter telescope and about a dozen other large telescopes. A branch of the observatory a few miles away on a peak called Cerro Pachón has a facility with the 4-meter SOAR reflector and the 8-meter Gemini South reflector. It also will be the site of the Vera Rubin Observatory, an 8-meter wide-field survey telescope under construction on the 8,900-foot-tall mountain.

Charles Messier — A French comet hunter who lived from 1730 to 1817. He published a list of "pesky" objects that could be confused with the faint comets he pursued tirelessly. The Messier Catalog became recognized by observers as a list of the 109 best and brightest galaxies, emission nebulae, and star clusters visible in the night sky.

CERRO TOLOLO:
Three of the large telescopes atop Cerro Tololo in Chile appear in silhouette against the broad band of the Milky Way. Two of our galaxy's satellites—the Large and Small Magellanic Clouds—appear above the two domes on the right. CTIO/NSF'S NOIRLAB/ AURA/H. STOCKEBRAND

Collimation — The alignment of the optics in your telescope. When a telescope is collimated, it will provide the sharpest images possible. Refractors maintain their collimation very well; it is possible you will never need to collimate a refractor. Big reflectors, as in big Dobsonian or Newtonian telescopes, need to be checked every time you transport the telescope.

Dark adaptation — The process of letting your eye adapt to seeing in a dark (nighttime) environment; it usually takes at least 10 minutes to become fully dark adapted.

Dark sky — Nighttime conditions when you are in a rural location far from city lighting. Usually this means at a minimum the Milky Way is visible to the unaided eye if it is up at that time of year and that time of night.

Dark of the Moon — When the Moon is not visible from your location and not expected to rise during your observing session. At New Moon, everywhere is under dark-of-the-Moon conditions for a couple days before and a couple days after.

Dark-sky location — Usually a specific rural location where you know a suitable night sky is available.

Declination (Dec.) — The equivalent of latitude on Earth for objects in the sky. The degrees of arc that an object lies north or south of the celestial equator.

Deep-sky object (DSO) — Any astronomical object outside the solar system. Double and variable stars, open and globular star clusters, emission and planetary nebulae, and galaxies all qualify.

Dobsonian telescope — A simple, sturdy Newtonian reflector on an alt-azimuth (left-right/up-down) mount developed and popularized in the late 20th century by John Dobson of the San Francisco Sidewalk Astronomers.

EASY-80 — An astronomical target in this book that is easily visible in an 80mm wide-field refractor. Easily visible means you can see it directly, not by averted vision.

European Southern Observatory (ESO) — The pan-European astronomy organization; they own the Very Large Telescope and are building the Extremely Large Telescope, a design consisting of 798 segments yielding the equivalent of a single 128-foot-diameter mirror scheduled to go into operation in 2027 on Cerro Amazones, Chile (a 9,993-foot peak).

Exit pupil — The width of the beam of light leaving an eyepiece, equal to the size of your objective lens divided by the magnification. Don't use an eyepiece in your telescope that produces an exit pupil greater than 7mm in diameter (it would be too large for your eye's pupil to take in).

Eyepiece — The removable lens on a telescope you look through to see an image. By changing the eyepiece, you can change the magnification and field of view of the telescope. In the United States, better telescopes usually have eyepieces 1.25 or 2 inches in diameter (the standard sizes commonly available). Cheaper telescopes may have 0.96-inch eyepieces (stay away from these; selection and quality is poor compared with the larger-sized eyepieces).

Field of view (FOV) formula: — FOV equals the apparent field of view (AFOV) of the eyepiece divided by the magnification of the telescope when using that particular eyepiece. For the wide-field 80mm refractor with a 68° 22mm eyepiece used in this book, the AFOV equals 68 divided by 23 (the power that eyepiece gives with the telescope), or 2.96°.

Gum Catalog — Similar to the Sharpless Catalog, this is a catalog of 84 emission nebulae for the southern sky, developed by the Australian astronomer Colin Stanley Gum at Mt. Stromlo Observatory in the Australian Capital Territory.

Herschel — Frederick William Herschel (1738–1822), Caroline Herschel (1750–1848), William Herschel's younger sister, and John Herschel (1792–1871). The Herschels were a famous family of astronomers, musical composers, telescope makers, and deep-sky observers who compiled what is now the NGC catalog and discovered many of the objects in this book. Roughly equivalent to the Kardashians in the 18th and 19th centuries (but the Herschels made things and discovered thousands of DSOs) or Illuminati during the Age of Enlightenment.

EUROPEAN SOUTHERN OBSERVATORY:
The European Southern Observatory runs many facilities in the Southern Hemisphere. Here, the telescopes of La Silla Observatory in northern Chile bask in the glow of the July 2, 2019, total solar eclipse. The eclipse brought nearly two minutes of totality to observers there. ESO/R. LUCCHESI

KITT PEAK NATIONAL OBSERVATORY:
The domes of Kitt Peak National Observatory in Arizona stand watch under the circular trails of stars in this time exposure.
The red glow on the mountaintop comes from lights designed to keep the astronomers working there dark adapted.

KPNO/NOIRLAB/NSF/AURA/B. TAFRESHI

High power — Produced by eyepieces with smaller focal lengths (approximately 4mm to 10mm of focal length). Power is equivalent to telescope magnification and equals telescope focal length divided by eyepiece focal length. For example, a telescope with 1000mm of focal length used with a 10mm eyepiece provides 100x or 100 power.

IC (Index Catalog) — The Index Catalogs were supplements to the NGC, published by Danish astronomer L.E. Dreyer.

Isaac Newton — Sort of the Elon Musk of the late 1600s and early 1700s; a famous mathematician and physicist, and a poster boy for the Age of Enlightenment. Invented calculus and the laws of gravity and Newtonian mechanics, among other accomplishments.

Kitt Peak National Observatory — A mountaintop observatory on Kitt Peak, a 6,886-foot-tall mountain outside Tucson, Arizona, run by National Science Foundation.

Light pollution — The brightening of the night sky caused by artificial lighting. Light pollution makes it hard to see the Milky Way or do deep-sky observing, interferes with sleep cycles, makes birds fly into buildings at night, and, according to some research, is related to some forms of cancer.

Light-pollution filter — Essential for observing emission and planetary nebulae. They reduce light pollution by restricting the transmission of all light except specific frequencies emitted by these nebulae. Most common are OIII (transmits light of doubly ionized oxygen), UHC or ultra-high-contrast filters (light of ionized oxygen and hydrogen), and H-beta (blue-green light of hydrogen).

Low power — Referring to eyepieces with larger focal lengths (approximately 20mm to 40mm). Power is equivalent to telescope magnification and equals telescope focal length divided by eyepiece focal length. For example, a telescope with 1000mm of focal length used with a 40mm eyepiece provides 25x or 25 power.

Magnification formula (recipe) — The focal length of a telescope divided by the focal length of the eyepiece or, as a formula, magnification (power) equals telescope focal length divided by eyepiece focal length (the focal lengths need to be in the same units, millimeters or inches).

Magnitude — A brightness scale for astronomical objects. The smaller the number, the brighter the object appears in the sky. The brightest stars are 1st magnitude and the faintest visible to the unaided eye are 6th magnitude. The difference of five magnitudes corresponds to 100 times in brightness. When you measure brightness exactly, it turns out a few stars have magnitudes of 0 or even a negative number, and the brightest planets, the Moon, and the Sun, have negative magnitudes. The faintest objects covered in this book are around 10th magnitude, or 100 times fainter than a 5th-magnitude star.

Magnitude limit — The faintest star the naked eye or a telescope can see. Larger telescopes can gather more light, so they have a fainter limit (in other words, they can show fainter stars, those with a larger magnitude number). An 80mm refractor used at 22x has a magnitude limit of about 11.8, but in practice, anything non-stellar fainter than magnitude 10.5 is difficult to see. Many factors, including an individual's eyesight, affect this limit. The limiting magnitude for a 6-inch refractor is about 13.4, the limiting magnitude for a 14-inch Schmidt-Cassegrain is about 15.3. You can find an online calculator at: https://www.cruxis.com/scope/limitingmagnitude.htm.

Mauna Kea — The site of the University of California/Caltech/NASA 10-meter Keck telescopes is a 13,796-foot-tall, long-dormant volcano in Hawaii where seven countries have built large telescopes. Astronomers regard it as one of, if not the, best observatory sites in the world.

MAUNA KEA:
Several large telescopes, including the twin 10-meter Keck instruments, call Mauna Kea in Hawaii home. The long-dormant volcano boasts some of the best skies on Earth.

INTERNATIONAL GEMINI OBSERVATORY/NOIRLAB/NSF/AURA/A. HARA

Meridian — The imaginary line that you could trace from true north to true south while passing directly overhead. As the stars cross the sky during the night, when they are on the meridian you will be looking through the least amount of atmosphere possible.

Objects are at their brightest and clearest when on or near the meridian.

Messier Catalog — The 109 bright deep-sky objects cataloged by Charles Messier (the "Ferret of Comets," a nickname given to him by King Louis XV of France). The first version was published in 1774.

New General Catalog (NGC) — A list of more than 7,500 non-stellar objects (star clusters, planetary nebulae, emission nebulae, and galaxies).

Newtonian — A reflecting telescope design invented by Isaac Newton (1643–1727). It is the optical design used in Dobsonian telescopes.

NOIRLab — National Optical and Infrared Astronomy Research Laboratory of the U.S. National Science Foundation. It has facilities on Kitt Peak, Arizona; Mauna Kea, Hawaii; Cerro Tololo, Chile; and Cerro Pachón, Chile.

Observe/observing — The activity performed by deep-sky observers when they search out and view deep-sky objects. This can be done anywhere, but more frequently it's associated with a trip to a dark-sky site and usually involves chasing down objects with a telescope. Still, it can be performed with the naked eye and binoculars.

Observer — A lover of the stars. An individual that enjoys viewing or simply looking at the cosmos; someone who in the context of deep-sky observing will be dragging a telescope to a dark-sky site to do some "observing" of astronomical objects.

Ocular — Another name for eyepiece.

Paranal — An 8,645-foot-tall mountain in Chile that's home to the Very Large Telescope (VLT). It is owned and operated by the European Southern Observatory. The VLT comprises four 8-meter telescopes that can operate independently or be combined into one telescope via interferometry.

Power — The magnification of your telescope, which depends on the eyepiece used; 25 power is shown as 25x.

Reflecting telescope — — Invented in 1666 by Isaac Newton (1643–1727), an English physicist, mathematician, and astronomer who invented the laws of gravity and calculus, among many other things. This type of telescope uses mirrors to collect and focus light. All large modern professional telescopes are reflectors.

Refracting telescope —The first type of telescope, invented by Dutch spectacle makers in the early

PARANAL OBSERVATORY:
The four 8-meter telescopes of the European Southern Observatory's Very Large Telescope (seen here at sunset) can operate either separately or as a unit. Working together, the quartet gathers as much light as a single 16-meter telescope would. ESO/F. KAMPHUES

1600s but turned to astronomical pursuits by Galileo Galilei (1564–1642). Refractors gather, focus, and magnify light by using lenses only.

Rich-field telescope — Short (low number) f/ratio telescope, usually f/4 to f/6; also called a wide-field telescope.

Right Ascension (R.A.) — The equivalent of longitude for objects in the sky. Measured in hours, minutes, and seconds. There are 24 hours of right ascension in a complete circle.

RNGC/IC — The Revised New General/Index Catalog, the latest revision of the New General Catalog, last published and updated in 2019 with 13,957 objects.

Schmidt-Cassegrain Telescope (SCT) — A telescope design that uses lenses and mirrors to result in a short and thus fairly transportable telescope tube.

Seeing — The stability of the air you are viewing through during the night. The steadier the seeing, the sharper the view you will get through a telescope.

Sharpless Catalog — A catalog of 312 HII regions north of declination –27° created by American astronomer Stuart Sharpless in 1953 (the second and final edition). For example, the Sharpless designation for the 54th object in his catalog, the nebula around NGC 6604 in Serpens, is Sh2-54 (the "2" designates that this is from the second and final edition).

Star-hopping — Finding deep-sky objects (DSOs) by moving from star to star (hopping) on a star chart until you can approximate where to point a telescope to see the object you're looking for. Aiming is then accomplished by pointing your telescope with an optical finder/red-dot to where you believe an object resides.

Supernova remnant (SNR) — The leftovers of an old star that has burned through its available supply of fusion fuel and explodes at the end of its life, creating a neutron star (pulsar) or a black hole along with a glowing gas cloud that remains visible until it dissipates.

Transparency — The clarity of the atmosphere. Transparency is lowered when smoke, dust, humidity, light, or other pollution makes it harder to see a deep-sky object. If the sky is very blue during the day, the transparency is generally good.

Uppsala General Catalog (UGC) — A catalog of 12,921 galaxies in the northern half of the sky brighter than magnitude 14.5 and first published in 1973 by the Uppsala Astronomical Observatory in Sweden. No person on Earth has visually seen all these objects through a telescope.

Zenith — The point in the sky exactly overhead. Observers hunt for objects when they are as close to the zenith as possible because that is where you are looking into space through the smallest amount of atmosphere possible.

Zwicky Catalog/CGCG Catalog — A catalog of galaxies and galaxy clusters compiled from 1961 to 1968 by Caltech astronomer Fritz Zwicky (a feisty contrarian who uncovered some of the first evidence for dark matter while observing from Palomar Mountain Observatory). It identifies 29,418 galaxies and 9,134 galaxy clusters.

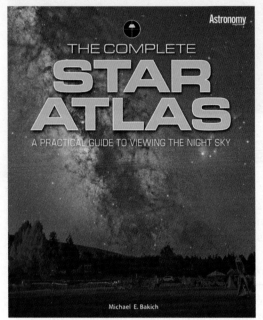